GEOLOGY OF THE JURASSIC COAST

The Red Coast Revealed

EXMOUTH TO LYME REGIS

Richard A Edwards

COASTAL PUBLISHING

A Coastal Publishing Limited Book

Author Dr Richard A Edwards
Project Editor Professor Denys Brunsden
Design Jonathan Lewis
Production Peter Sills

First Published 2008
Reprinted 2013 with minor amendments
Coastal Publishing Limited
The Studio, Puddletown Road, Wareham, Dorset BH20 6AE
Tel: 01929 554195
email: enquiries@coastalpublishing.co.uk
www.coastalpublishing.co.uk

ISBN 978-0-9544845-4-5

British Library Cataloguing-in-Publication Data
A catalogue record for this book is available from the British Library.

Your purchase of this book helps to support the conservation and education programmes of the Jurassic Coast Trust (Registered Charity No: 1101134).
www.jurassiccoast.org/get-involved

Exmouth Town Council, Natural England and Devon County Council are proud to be associated with the Jurassic Coast Trust and have contributed financially to the origination of this publication.

Front cover image: Ladram Rocks

Printed and bound in the United Kingdom.

For Jenny and Katie

Richard Edwards

Graduated from Nottingham University and then moved to Exeter University where he worked on the joint University/ Geological Survey venture to resurvey the Newton Abbot geological map sheet, and completed his PhD thesis on the geology of the Bovey Basin. Subsequently he was for many years a professional geologist with the Geological Survey, firstly in Africa (Zambia) and later in Hampshire, Devon and Somerset. He was involved in the detailed geological surveys of the 1:50,000 - scale Newton Abbot, Southampton, Exeter, Minehead and Sidmouth map areas. He is the author of 'Exmoor Geology', a popular account of the rocks and landscapes of Exmoor National Park (2000), and 'Devon's Non-Metal Mines', a description of the county's slate, culm, whetstone, Beer Stone, ball clay and lignite mines (2011).

JURASSIC COAST Trust

The Jurassic Coast Trust

The Jurassic Coast Trust, established in 2002, is an independent registered charity governed by a board of trustees, which supports the delivery of science, conservation, arts, education and sustainable development programmes.

The Trust provides advice and influence and works with the Jurassic Coast World Heritage Team and other partners.

If you would like to support the work of the Trust please visit www.jurassiccoast.org/get-involved

Contents

Introduction

The story of the rocks and landscapes

A journey through the rocks and landscapes: Exmouth to Pinhay Bay

Useful stone and minerals

"Here, in the middle of the night, a tract of ten acres...suddenly sank down from 200 to 260 feet, presenting a striking group of shattered pinnacles and columns of chalk intermingled with the sunken fragments of the fields thus torn away from their native site; the remains of hedges still traversed these fragments, and a stile was seen undisturbed on the summit of one of the subsided columnar masses. The eye looking from above on this wild scene of ruin, saw it finely backed by the deep blue of the channel beneath; for in this case the landslip in question formed the first term in the commencement of a new and incipient undercliff..."

A vivid description of the 1790 Hooken Landslip between Branscombe and Beer, by the Victorian geologists Conybeare and Buckland, written in 1840.

This book is dedicated to the memory of the celebrated geologist W A E Ussher (1849-1920), who made remarkable contributions to the understanding of the geology of SW England, including this part of the Jurassic Coast.

Travelling the 155 kilometres of foreshore that twists and turns from Exmouth to Lyme Regis is not simply a journey through some of the most sublime coastal scenery in Britain, it is also time travel of the most fabulous kind. In the hour or so that it can take to drive from one end of this World Heritage coast to the other, you dip in and out of 265 million years of Earth history, not to mention enjoying the opportunity of some timeless pub lunches and cream teas.

The keys to unravelling this magnificent geological legacy come from the rocks underfoot, which under the expert tuition of this book become windows into truly amazing lost worlds. Walk the red pebble cliffs of east Devon, for example, and you become transported back to flash flood wadis in a vast desert when southern England lay at the heart of a great amalgam of land masses - the supercontinent of Pangaea. Alternatively, pick your way gingerly through the landslipped detritus of Charmouth and you find yourself among the remnants of Jurassic seas, a voyeur on an ancient sea bed at a time when the world's oceans burst into diverse life. Or descend the narrow streets of Beer to encounter the formidable chalk cliffs that, further east, rise into our nation's southern ramparts, solid buttresses built on the unlikely shoulders of tiny algae. Whichever nook or cranny you choose to explore here, there are joys to behold.

The remarkable rocks that stretch out along the Jurassic Coast are the training ground for many a young or budding geologist. It was here that I cut my early professional teeth, leading annual pilgrimages of London University undergraduates westwards on field courses to this classic piece of British geology. Now, whether I'm on a journey around some ring of fire or to some distant corner of the planet, many of the geological reference points are tied to the bays, headlands and sea stacks of east Devon and Dorset. It helps that, having moved to Plymouth, this wonderland is now on my doorstep.

But if you think 'The Red Coast Revealed' is a guide only for professional geologists and committed rockhounds, think again. Amidst the lucid explanations that carefully and patiently reveal the landscape and the rocks, are scattered wonderful human stories. Derivations of local place names jostle with historical footnotes and anecdotal asides in celebrating how geology provides the foundations for the architectural and cultural riches of this distinctive region. It all makes for a lively and fascinating read. For those embarking along this very special piece of coast, this isn't so much a geological field guide but, rather, an essential travelling companion.

Dr. Iain Stewart
Geologist and television presenter

The Hooken Landslip: in 1790 huge masses of Chalk and Greensand slid down between Branscombe and Beer to produce the present tumbled landscape.

The 'Jurassic Coast'

Orcombe Point at the western end of the World Heritage Site is marked by the Geoneedle, a sculpture made of rocks from the various formations of the Site and opened by HRH Prince Charles in 2002.

On 13th December 2001, the whole 155km length of coast between Exmouth (Orcombe Point) in Devon and Old Harry Rocks in Purbeck, Dorset, was designated a World Heritage Site by the United Nations, immediately and excitingly putting it in the same league as the Grand Canyon and other famous World Heritage Sites. The designation was a recognition primarily of the international importance of the geology and landforms of the coastline. This stretch of coast is important because it displays a unique, nearly continuous, sequence of rocks recording 265 million years of earth history. The rocks belong to the geological periods of time called the Permian, Triassic, Jurassic, Cretaceous, Tertiary and Quaternary and represent a wide range of past environments. They are superbly displayed in the cliffs of the area, since they are mostly tilted gently eastwards so that, travelling from west to east, we encounter progressively younger rock strata, from the oldest (Permian) in the west to the youngest (Cretaceous) in the east. To cap it all, the whole sequence forms an easily accessible and unspoilt coastline of great scenic beauty.

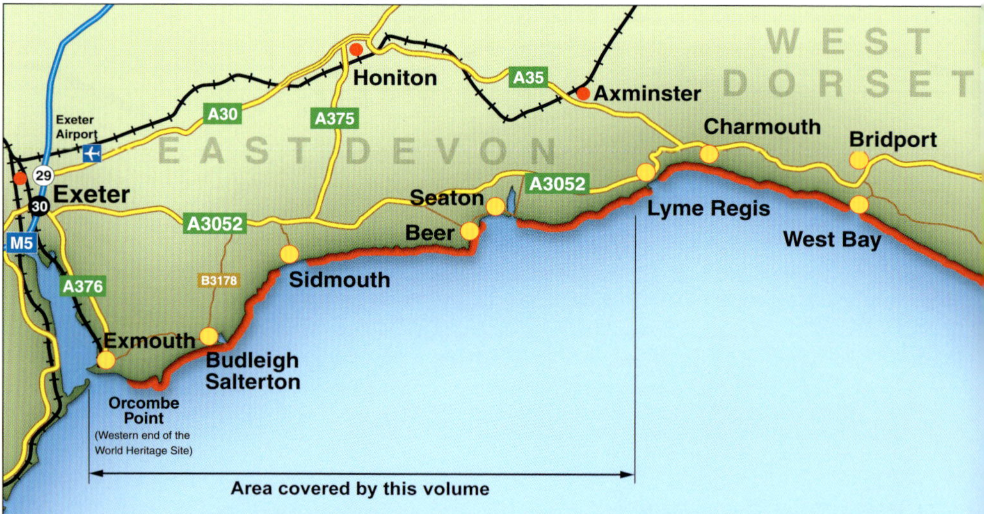

SAFETY FIRST!

Rockfalls and mudslides are an ever-present hazard on this coast and you are most strongly advised to stay away from the base of the cliffs and the cliff top. See pages 126-127 for more information about safety and fossil collecting.

In addition to the geological significance of the coast, its landforms are also of great interest and include examples of spectacular landslips, cliffs, shingle ridges and saltmarshes.

Old Harry Rocks mark the eastern end of the World Heritage Site. Natural sculptures of great fame made from Cretaceous Chalk, they are classic examples of sea-stacks.

The designation of the World Heritage Site opens up great educational and visitor opportunities for the explanation of its earth history. The World Heritage Site coast has already been informally named the 'Jurassic Coast' after the best known of the geological periods represented along it.

As well as its recent inclusion in the World Heritage Site, the scenic value of our particular area has also been recognised in its existing designation as part of the East Devon Area of Outstanding Natural Beauty (AONB). The coast is part of the East Devon Heritage Coast, and large stretches are also designated as Sites of Special Scientific Interest. The National Trust owns considerable parts of the coast.

About this book – the 'Red Coast'

This book focusses on the part of the World Heritage Coast within Devon, from Exmouth in the west to Pinhay Bay (close to the Devon-Dorset border) in the east. In between lie the charming towns and villages of Budleigh Salterton, Sidmouth, Salcombe Regis, Branscombe, Beer and Seaton. This part of the coast is composed mostly of Permian and Triassic rocks (overlain in places by Cretaceous rocks) and might therefore be named informally the 'Red Coast' since most of the rocks are that colour. In the area covered by this book, Jurassic rocks only appear east of Seaton, and extend eastwards from there to the famous cliffs of Lyme Regis. I have also described, for completeness, the coast between Exmouth and Orcombe Point which, although outside the World Heritage Site, has interesting features.

My aim is to give you a gentle and enjoyable introduction to the scenery and rocks of this beautiful part of the Devon coast. I will explain how the story of the rocks that are so spectacularly displayed in the cliffs can be read to give an insight into their earth history. We shall also see how the geological foundations of the area have influenced the shape of today's landscape. I will, in effect, take you on a journey through the earth history of our area from about 265 million years ago to the present day. I have also included something of man's influence on the landscape, either through quarrying for useful stone or minerals, or the construction of prehistoric hillforts. Whether you live in the area or are one of our many welcome visitors, I hope that all those of you with a curiosity in your natural surroundings will be able to dip into the book and find something of interest.

The book is easy to use, since you don't need any special knowledge of geology to read it - just an interest and curiosity in your natural surroundings. It's the sort of book that you might use while on holiday or out for a day trip, and which might add some extra interest to the walks or excursions that you have planned. I have tried not to include too many unfamiliar geological terms – those that are used are explained in the glossary at the end of the book.

Before we start – a brief word about safety! The cliffs are being actively worn away in many places and falls can occur at any time – please be aware of these and other hazards and read carefully the safety advice on pages 126-127.

An introduction to the rocks of the Red Coast

There are several very clear broad divisions of the rocks of the Red Coast that are beautifully revealed in the cliffs. The coastline is dominated by cliffs of red mudstones and sandstones which belong mainly to the Permian and Triassic periods of earth history. They are capped east of Sidmouth by paler coloured sandstones and chalk belonging to the Cretaceous period. East of Seaton, towards Lyme Regis, the Triassic rocks are succeeded by grey rocks belonging to the Jurassic period, including the famous Blue Lias. The rock formations of our area are listed in the table on page 13, and their distribution is shown in the fold-out geological map at the back of this book. Cross-sections through the rocks along the coastline between Exmouth and Lyme Regis are shown in the fold-out section also at the back of this book.

Aerial view showing the mouth of the River Otter near Budleigh Salterton. The river flows through a narrow gap in the shingle ridge. The red cliffs east of the river mouth are of Triassic Otter Sandstone.

The history of the landscapes and visible rocks goes back to a time about 265 million years ago, during the Permian period. Permian rocks occur along the coast between Exmouth and Budleigh Salterton. In Exmouth they include a type of ancient gravel called breccia, and named the Exe Breccia. The breccias are succeeded eastwards by red mudstones with some sandstones, called the Aylesbeare Mudstone (about 400m thick). They form the cliffs between Orcombe Point and Budleigh Salterton.

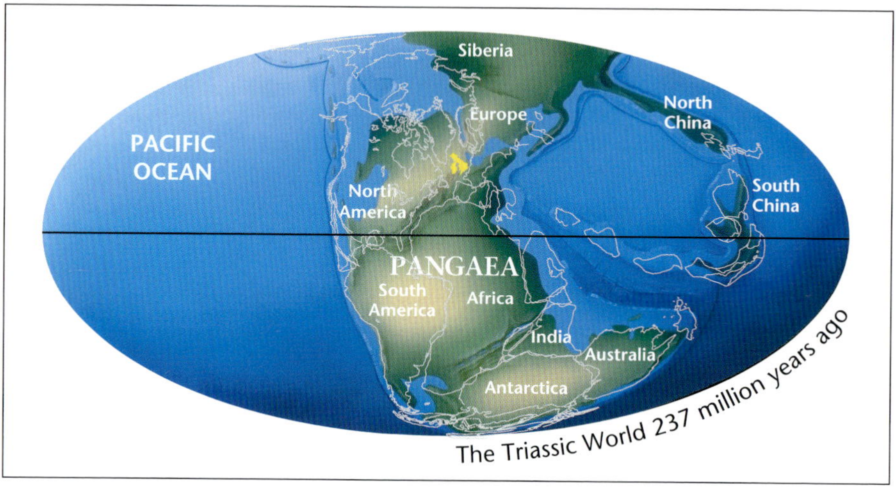

The following Triassic period spans the period of earth history between 250 and 200 million years ago. During the Triassic the geography of the world was totally different to that of today. Then, all the continents were joined together in a super-continent called Pangaea (above). Its centre was occupied by deserts, with greener areas around the coasts or along rivers. Animal life shows the diversification that followed the greatest of the mass extinctions of living things which happened at the end of the preceding Permian period (about 250 million years ago). The Triassic animals included primitive amphibians, rhynchosaurs ('beaked lizards') and primitive kinds of crocodiles. The dinosaurs appeared on the scene about 230 million years ago, at the beginning of the Late Triassic. Flowering plants had not yet evolved in the Triassic, and the only plants were non-flowering kinds, such as ferns, conifers, cycads and gingkos.

Triassic rocks occur in the cliffs between Budleigh Salterton and Branscombe - They reappear east of Beer at Seaton Hole.

The Triassic rocks record a time when Devon lay only about 10 degrees north of the equator, and the area was a hot desert. Although the rocks are all red in colour and so superficially similar, they actually include several quite different

Aeonium arboreum: a direct descendant of plants from the Triassic Period.

Period	Name	Subdivision	
Quaternary	Shingle ridges, saltmarsh deposits, river terrace gravels, alluvium, peat, head and landslips		
Tertiary & Quaternary	Clay-with-flints		
Cretaceous	Chalk	'Upper Chalk'	
		'Middle Chalk'	
		Beer Head Limestone	
	Upper Greensand	Bindon Sandstone	
		Whitecliff Chert	
		Foxmould	
	Gault		
Jurassic	Charmouth Mudstone		
		Shales-with-Beef	
	Blue Lias		
Triassic	Penarth Group	Lilstock Formation	Langport Member ('White Lias')
			Cotham Member
		Westbury Formation	
	Mercia Mudstone	Blue Anchor Formation	
		Other formations, including Dunscombe Mudstone Formation	
	Otter Sandstone		
	Budleigh Salterton Pebble Beds		
Permian	Aylesbeare Mudstone	Littleham Mudstone	
		Exmouth Mudstone and Sandstone	
	Exe Breccia		

Rock formations of the Red Coast, their periods, names and subdivisions.

types, ranging from mudstones, through sandstones to conglomerates. These red rocks were originally laid down as flat layers (beds), but subsequent earth movements have given them an eastward tilt. This has the effect of introducing successively younger red rocks as the cliffs are followed from west to east.

The Triassic Budleigh Salterton Pebble Beds form a layer of pebbles about 30m thick in the cliffs west of Budleigh Salterton. They dip eastwards at about 5 degrees and thus disappear beneath the next overlying layer, which consists of sandstones called the Otter Sandstone, of Mid Triassic age, up to 210m thick. These form the cliffs between the mouth of the River Otter and High Peak, near Sidmouth, where they in their turn dip eastwards down to beach level to disappear beneath the next youngest layer, the Triassic Mercia Mudstone. These mudstones (totalling about 440m thick) form the lower cliffs between High Peak and Sidmouth and most of the lower cliffs between Sidmouth and Branscombe Mouth.

In the higher part of the cliffs between Sidmouth and Branscombe Mouth, the Mercia Mudstone is overlain by a generally flat-based series of much younger sandstones and chalk (the Upper Greensand and overlying Chalk). These belong to the Cretaceous period, between 145 and 66 million years ago. In contrast to the Triassic period, the arrangement of the continents in Cretaceous times was not too different from today, although a large ocean lay between 'Europe' and 'Africa'. The Cretaceous rocks record the establishment of conditions totally different from those of the red rocks, and were deposited on the bed of a sea which spread onto and drowned the land area. Such a process is called a marine transgression. The flat-lying Cretaceous rocks cut across the inclined layers of the red rocks, forming a classical example of an angular unconformity.

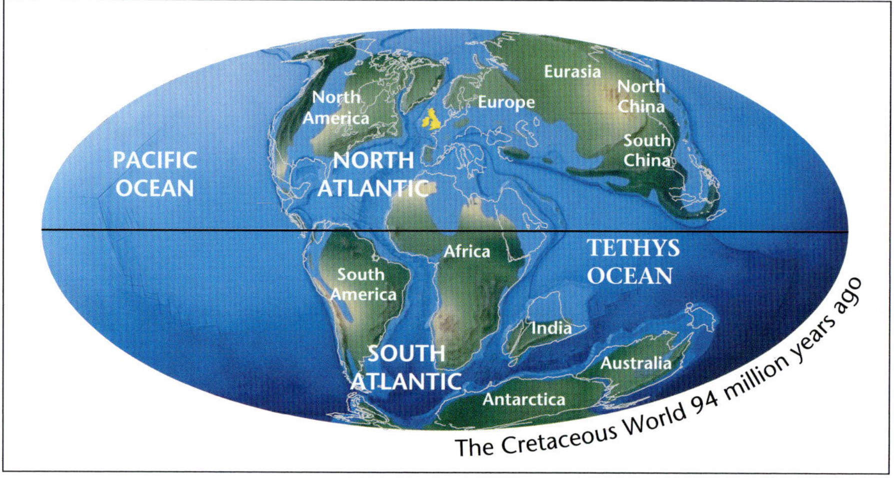

The Cretaceous World 94 million years ago

At Sidmouth, a crack or fault at Jacob's Ladder brings the Otter Sandstone up from the depths, and it extends eastwards beneath Sidmouth and into the cliffs east of the mouth of the River Sid. There, as below High Peak, the Otter Sandstone again dips eastwards beneath the Mercia Mudstone and is not seen again. The Mercia Mudstone, still eastward dipping, forms the lower part of the cliffs all the way to Branscombe Mouth. The Upper Greensand, together now with the Chalk, forms the higher part of the cliffs all the way to Branscombe Mouth. East of there, the Cretaceous rocks, until now roughly horizontal, are tilted eastwards and come to beach level at Beer Head

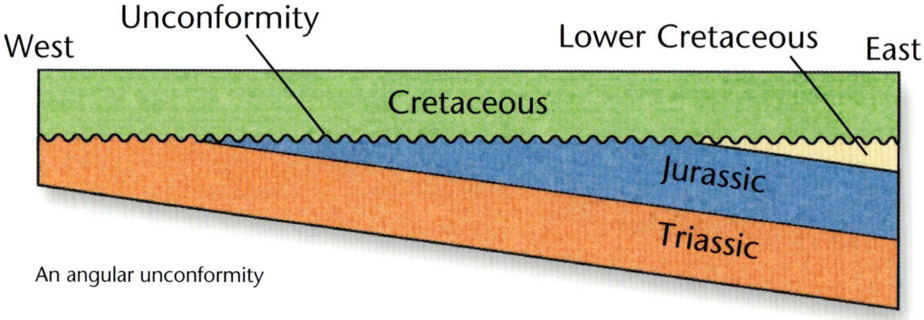

An angular unconformity

The Chalk and Upper Greensand are tilted eastwards and come to beach level at Beer Head in the distance. An unconformity is evident where the Cretaceous rocks overlie the Triassic Mercia Mudstone. The Jurassic rocks between are missing owing to erosion about 120 million years ago.

Sir Henry De la Beche (1796 -1855), founder of the Geological Survey.

and then form the cliffs round the coastline into Beer Roads. At Seaton Hole, between Beer and Seaton, there is another major fault which brings the Mercia Mudstone up from the depths to form the cliffs between Whitecliff and Seaton. Between Seaton Hole and Seaton, the low cliffs are made up of red Mercia Mudstone with some small faults.

East of the mouth of the River Axe, towards Pinhay Bay, the coastline is extensively landslipped, but in places the cliffs expose some of the formations. Here we see the transition from the Triassic rocks up into the Jurassic rocks. The upper part of the Triassic Mercia Mudstone, including the Blue Anchor Formation, is visible, and farther east the Triassic Penarth Group (including a distinctive limestone, the Langport Member or 'White Lias') and the Jurassic Blue Lias and Shales-with-Beef (belonging to the lower part of the Charmouth Mudstone) appear. The upper cliff is made up of Upper Greensand and Chalk, which have been involved in great landslips.

The flat-topped Greensand and Chalk hills are capped by a layer of brown clay with unworn flints called the Clay-with-flints. It is normally a layer up to about 5m thick, but in places fills deep solution pipes descending to considerable depths into the Chalk or Upper Greensand.

During the last two million years or so, a period of earth history known as the Quaternary, a variety of deposits formed, and some, such as beach deposits, are still forming today. They include: deposits formed by the sea or in the intertidal zone, including modern beaches and shingle ridges and also saltmarsh deposits; deposits formed by rivers, including river terrace gravels and modern alluvium; and deposits formed by mass movement processes – mainly head and landslips.

The rocks of the Red Coast are summarised in the table on page 13. The fold-out map at the back of this book also shows the geology with the names of the main places of interest.

Early geological studies

Because the geology is so clearly laid out in the cliffs, the coast has attracted the attention of geologists since the earliest days of the science. They include some of the great names of early geology such as Sir Henry De la Beche (the founder of the Geological Survey), the Rev William Buckland and the Rev W D Conybeare. As early as 1819, De la Beche read a paper for the Geological Society which included a pictorial section showing the geology of the cliffs.

William Buckland
(1784 - 1856)

Buckland discussed early controversies between science and religion in an 1822 paper 'On the Excavation of Valleys'. He noted that the hills and valleys of East Devon are abruptly terminated by the sea, and considered that the present-day streams were incapable of having shaped the valleys. He concluded that the valleys had been eroded by torrential floodwaters which he took as evidence of the biblical Flood. In his Reliqiae Diluvianae ('Relics of the Flood'), of 1823, there is an excellent sketch of the coast section between Sidmouth and Beer Head (see pages 38-39).

P O Hutchinson
(1810 - 1897)

The great Bindon Landslip, east of Seaton, which occurred on Christmas Day 1839, was the first in Britain to be extensively documented and analysed by the early geologists.

The well-known Sidmouth antiquarian, P O (Orlando) Hutchinson, had a great interest in geology, and in 1843 produced a guide to the geology of the Sidmouth area.

W A E Ussher
(1849 - 1920)

The first geological map of the area, on a scale of one-inch to the mile, was made by De la Beche, and published in 1834. The area was mapped by the Geological Survey, again on the one-inch scale, between 1873-1887, and one of the surveyors was W A E Ussher, a remarkable man who had a great influence on the geological understanding of S W England. The two modern geological maps covering the area are the 1:50,000-scale sheets 339 (Newton Abbot), issued in 1976, and 326 (Sidmouth), issued in 2005.

What maps to use?

The geological sheets 339 and 326, mentioned above, show the detailed geology. As far as topographical (Ordnance Survey) maps are concerned, the 1:25,000 (2½ inches to 1 mile) Explorer Maps are indispensable. The area is covered by two sheets: Explorer 115 (Exmouth & Sidmouth) and Explorer 116 (Lyme Regis & Bridport). The 1:50,000 Landranger maps 192 (Exeter, Sidmouth & surrounding area) and 193 (Taunton & Lyme Regis) are also useful.

Geological terms are explained in the Glossary, pages 116-122. Figures in square brackets are National Grid references for the locality indicated. They lie within the 100km square SY. See the margins of Ordnance Survey maps for guidance on how to use the National Grid system.

THE STORY OF THE ROCKS AND LANDSCAPES

The cliffs west of Beer Head show masses of Chalk and Upper Greensand forming part of the Hooken Landslip.

Before we look at the story told by the rocks, it is worth a quick, and hopefully painless, look at some of the basic concepts of geology. This will help us to appreciate and understand more readily some of the features that we will encounter later when we look at the rocks and landscapes in more detail.

Geology – the story of the earth

When we walk down a street, or across a grassy field, few of us give thought to what lies beneath our feet. Yet concealed below only a thin layer of tarmac or soil lies the world of the rocks – hidden from view except where quarries have been dug, or rivers have cut deeply down, or the sea has formed cliffs. These rocks have a rich and varied story to tell, if we can read the clues in them, and the study of them makes up the science of geology. The fascination of geology is that we can understand the past story of our planet, written in the rocks. We can also understand the origin of our modern landscapes that are modelled from the underlying rocks. The World Heritage Coast is a particularly good area in which to study the close relationship between the geology and the landscapes to which it gives rise.

Practical geology

Geology is not just interesting for its own sake – it is hugely relevant to our everyday life. One very important aspect of the science is the search by geologists for natural resources, like oil, gas, coal and metal ores, as well as bulk minerals such as sand, gravel and clay - all vital to the existence of modern society. This more traditional role of geology has been added to in recent years by that of environmental protection, for example, in such important questions as the risk of flooding, the disposal of radioactive waste, the prediction of earthquakes and volcanic eruptions and the factors leading to rising sea levels and coastal erosion.

The cliffs between Orcombe Point (nearest) and Straight Point (in the distance) are made up of eastward-dipping red mudstones with some sandstones, part of the Aylesbeare Mudstone. Harder more resistant beds form ridges along the foreshore.

The vastness of geological time – the geological time scale

The earth is incredibly ancient - possibly over 4.6 thousand million years old. A great division in the history of the earth occurred 541 million years ago. Rocks older than that date, which represent the vast bulk of earth history, are called Precambrian (see the table below), and contain few fossils. About 541 million years ago, there was a remarkable 'explosion' in the number of easily preserved shelly animals, now found as fossils in the rocks. The last 541 million years is divided by geologists into parcels of time with special names. This geological time scale, with the named periods of time, is shown in the table. The periods are sometimes called after places where the rocks were first studied or are best displayed (a familiar example is the Jurassic – the 'age of the dinosaurs' - named after the Jura Mountains in France). Sometimes the periods are named after special characters of the rocks (for example, the Carboniferous, from the fact that the rocks contain coal [carbon] seams). The Permian, Triassic and Cretaceous are of special significance within the Red Coast, since most of the rocks belong to those three periods. The Permian was named from the province of Perm, in Russia, where rocks of this age were first studied. The Triassic is named after the three-part division of rocks of this age in Germany. The Cretaceous is named after the Latin word for chalk ('*creta*') because of the British Chalk beds of this age.

The precise dates calculated for the beginnings and ends of the various geological periods tend to change as geologists carry out more detailed research to refine the time scale. The dates in other books may differ from those shown below. The ages

Era	Period	Epoch	Age (Millions of Years)	First appearance of:	Major events
	Quaternary	Holocene	.01		
		Pleistocene	2.6	*Homo erectus*	Ice Ages in northern hemisphere
Cainozoic	Neogene	Pliocene	5.3		
		Miocene	23.0	Apes	Alpine mountains
	Palaeogene	Oligocene	33.9	Monkeys	
		Eocene	56.0	Horses	
		Palaeocene	66.0		Great extinction of life at end of Cretaceous
Mesozoic	Cretaceous		145.0	Modern mammals & flowering plants	Chalk widely deposited
	Jurassic		201.3	Birds & dinosaurs	Ammonites abundant. Dinosaurs on land
	Triassic		252.2		Deserts. Great extinction of life at end of Permian
Palaeozoic	Permian		298.9	Theraspids	Deserts. Pangaea supercontinent
	Carboniferous		358.9		Coal forms in swamps
	Devonian		419.2	Amphibians & sharks	Fish and amphibians
	Silurian		443.8	Ammonoids & jawed fish	Caledonian mountains
	Ordovican		485.4	Land plants	
	Cambrian		541.0	Trilobites	
Precambrian	Proterozoic		2500	Soft-bodied animals	Glaciations. Various periods of mountain building
	Archaean		About 4600		

Rocks found in the World Heritage Site

The geological time scale. From 'The geological time scale 2012' (Gradstein and others 2012).

used are based on the most recent research, collected in a review of the geological time scale in 2012 (see Gradstein, in the Further Reading section, on page 125).

The present is the key to the past - 'uniformitarianism'

A basic principle of geology is the 'principle of uniformitarianism'. It simply means that geological processes that operate at present also operated in the past and will continue to do so in the future (although the intensity of the process may vary). For example, a 200 million-year-old sandstone with ripples is assumed to have been formed in the same way as today – by the action of water or wind on loose sand. Similarly, rocks formed from molten rock in modern volcanic eruptions can be compared with similar-looking ancient rocks. It can be assumed that the ancient rocks also formed from volcanoes. Thus 'the present is the key to the past'.

The physical landscape seems to be permanent and unchanging. This is because the geological perspective is much longer than the tiny length of the human life-span. In reality, the physical world is in the course of constant change, mostly very gradual. The huge spans of geological time available for change are crucial: even gradual processes can dramatically change the shape of the land. Not all processes act smoothly and slowly. Some geological events – such as earthquakes, meteorite impacts, volcanic eruptions and gigantic floods – are catastrophic and cause very rapid changes. Some catastrophic events, particularly large meteorite impacts, are so rare that we cannot use the present as a key to the past. For example, we have not - fortunately! - observed the arrival of a very large meteorite and thus have no modern example of its effects.

The three rock families

Rocks are divided into three families, depending on how they formed. They are igneous, metamorphic and sedimentary. The first two rock types are dealt with briefly, since they do not occur in our area.

Igneous rocks *(named from the Latin ignis, meaning 'fire')* form from hot molten material which cools to form solid rock. Examples are granite and volcanic rocks.

Metamorphic rocks *(named from the Greek meta 'change' and morphe 'form')* are rocks that have been changed from their original state by the action of heat or pressure, often deep beneath the earth's surface. Many Precambrian rocks are metamorphic.

Sedimentary rocks *(named from the Latin sedimentum 'settling')* form at the Earth's surface. They are derived from existing rocks of all three families by weathering and erosion, producing particles which are then laid down by water or wind in new areas. Generally they later undergo a hardening process in which the particles are cemented together. There are two classes of sedimentary rock: detrital (from the Latin for 'worn down') consisting of rocks made up of particles derived from the breakdown of pre-existing rocks; and chemical rocks formed by direct chemical precipitation, although, to make things difficult, the distinction is sometimes blurred!

All the rocks in our area belong to the sedimentary family. The main detrital types are:

Conglomerates consist of pebbles and cobbles cemented together. The main type in our area forms the Budleigh Salterton Pebble Beds, composed of rounded pebbles and cobbles in a sandy matrix. The Pebble Beds are strictly speaking gravel rather than conglomerate, since the pebbles are not usually cemented. Conglomerates also occur as thin beds in the Otter Sandstone. In this case they are true conglomerates since the pebbles are strongly cemented together to form hard layers. The pebbles are mainly well-rounded quartzite from the same source as the Pebble Beds. A small proportion, about 15%, is made up of 'foreign' pebbles transported into our area from a distant source. Other pebbles in the Otter Sandstone conglomerates are locally derived and consist of a mixture of calcrete, sandstone and mudstone fragments. Very occasionally, pieces of fossil vertebrate bone may occur.

Sandstones consist of sand grains cemented together. They make up most of the Otter Sandstone, a small part of the Budleigh Salterton Pebble Beds and also the Upper Greensand. A few thin beds of sandstone occur in the Mercia Mudstone, and also in the lower half of the Aylesbeare Mudstone (the Exmouth Mudstone and Sandstone). Sandstones of the Otter Sandstone consist of fine- to medium-sized grains of quartz and quartzite with some grains of the mineral feldspar. The sandstones are rather weakly cemented, compared to the better-cemented conglomerates (see left) so that the Otter Sandstone is not generally strong enough to use as a building stone. Cross-bedding is common in the Otter Sandstone. As the name suggests, it forms at an oblique angle to normal bedding, by avalanching of sand grains down a slope, such as the bar of a river. It can also form by wind action.

The Upper Greensand varies in character. West of the Sid Valley, it is mostly unconsolidated sand rather than sandstone. East of the Sid Valley, from Salcombe Hill eastwards, the Upper Greensand forms crags in the upper part of the cliffs and consists of calcareous sandstones and calcarenites (a type of sandstone with a very high (over 50%) content of calcareous detritus, cemented by calcite). It is possible that the lack of carbonate in the Upper Greensand west of the Sid Valley is linked to the removal of the former cover of Chalk, which allowed the calcareous cement to be leached away.

Breccias are like conglomerates, except that the pebbles and cobbles are angular. Breccias make up the Exe Breccia in Exmouth. Thin breccias also occur within the Otter Sandstone.

Mudstones consist of clay and silt particles cemented together. They make up most of the Aylesbeare Mudstone, Mercia Mudstone, Blue Lias and Charmouth Mudstone. Minor amounts occur in the Otter Sandstone. Most of the Permian and Triassic mudstones are reddish brown in colour, with a characteristic 'blocky weathering' habit. They used to be called 'marls' but actually do not contain enough carbonate to justify the name. Some bands of green mudstone occur within the red and can form persistent layers in the cliffs. Green patches and small green spots ('reduction spots') are also common. At Littleham Cove [SY 040 803], some reduction spots have dark radioactive centres with high concentrations of vanadium, uranium, copper and nickel.

In the middle of the Mercia Mudstone, around the Weston Mouth area, a layer of green and purple mudstone called the Dunscombe Mudstone can be distinguished from the mainly reddish brown mudstone above and below. The Blue Anchor Formation at the top of the Mercia Mudstone consists of grey and green mudstones. The mudstones of the Jurassic Blue Lias and Charmouth Mudstone are grey in colour. Black mudstones occur in the Penarth Group.

Limestones consist mainly of calcium carbonate. The most distinctive kind is chalk – a type of limestone made up fragments of calcareous algae (such as coccoliths) and other micro-organisms. Other types of limestone, such as the Beer Head Limestone at the base of the Chalk, are made up of broken fragments of shell. The Beer Stone in the Middle Chalk is an unusual type of limestone which includes much echinoderm debris. Limestone ('White Lias') is present at the top of the Penarth Group, and thin limestones are interbedded with mudstones of the Blue Lias.

The chemical types of sedimentary rock include:

Flint consists of nodules of very hard silica, commonly occurring as layers in chalk. Flint is a replacement of very fine-grained carbonate and consequently has a denser, finer-grained texture than chert.

Chert consists of nodules or tabular layers of very hard silica, that commonly occur as layers in the Upper Greensand, especially in the Whitecliff Chert. Chert is also, like flint, a replacement of original carbonate material, but in the case of chert the material is coarser grained calcarenite, sometimes with original fossils and sedimentary structures such as cross-bedding preserved.

Calcrete consists mainly of calcium carbonate. It is present in the Otter Sandstone as vertical tubular calcareous concretions and also as horizontal or inclined calcareous sheets.

Gypsum is hydrous calcium sulphate. It is a soft white or pink mineral that forms nodules or veins in the Mercia Mudstone. In places the gypsum has been dissolved away and only hollows remain.

Baryte is a heavy mineral, barium sulphate, that has been recorded from nodules beneath Peak Hill by Keith Corrie and David Green (see Further Reading, page 124-125). Most, if not all, of the specimens recorded as celestine (strontium sulphate) by earlier collectors are probably in fact baryte.

Silcrete consists of angular fragments of flint cemented together by silica. It is associated with the Clay-with-flints plateau.

Layered rocks – beds and bedding

One of the most distinctive features of sedimentary rocks is that they are laid down in individual episodes of deposition, each differing to a greater or lesser extent from its neighbouring layer in grain size or composition. These individual layers are called beds (or strata – singular stratum), and the layered character is called bedding. Good examples of bedding can be seen in the sandstone cliffs between Budleigh Salterton and Ladram Bay (see below). Another important point is that the sedimentary layers or beds are for the most part laid down horizontally. A famous example of horizontal layers is the Grand Canyon where the River Colorado has cut down for over 6000 feet into a series of horizontal beds ranging in age from Precambrian to Permian. As always, there are some exceptions to the principle of original horizontality – for example, some sedimentary rocks – such as sand dunes – originally accumulated on sloping surfaces.

Commonsense tells us that the lowest layer in an undisturbed sequence of layered rocks has to be older than the higher layers – a concept called the principle of the superposition of strata. This, together with the principle of uniformitarianism (see page 22) is one of the two basic 'laws' of geology. However, sometimes whole blocks of rock can be turned upside down by earth movements, in which case the highest layers become the oldest! Fortunately this complication does not apply to the Red Coast, where the rocks are either flat lying or gently tilted.

Earth movements and layered rocks – unconformities

Where layers of sedimentary rocks are tilted it can generally be assumed that earth movements have affected them. On the Red Coast there is a marked contrast in attitude between the red rocks of Triassic age which have been tilted eastwards by angles of about 5 degrees, and the Cretaceous rocks which mostly overlie the red rocks horizontally. This indicates a period of earth movements separating the formation of the two groups of rocks. An unconformity represents a gap in a rock

Cliffs of Otter Sandstone at Danger Point, just east of the mouth of the River Otter.

sequence. There is no place on earth where sedimentary rocks have been laid down continuously throughout geological time - there have always been interruptions due to earth movements. For example, an area which was once under the sea might be raised above sea level by earth movements to form land. During this time - which might last millions of years - there might be no rocks formed on the land surface. Then the land might be engulfed by the sea again, with renewed deposition of sediments. When preserved, this sequence would therefore show two periods of marine deposition separated by a large gap in deposition. The surface between the two sequences is an unconformity. In a case where the older sequence of rocks has been tilted by earth movements, the younger rocks would overlie the tilted sequence horizontally, and the surface between the two is called an angular unconformity. This is of particular relevance in East Devon because it is precisely the relationship which we find between the older Triassic red rocks and the younger Cretaceous rocks.

Peak Hill, west of Sidmouth, shows the unconformity between the yellow - brown Cretaceous (Upper Greensand) above and the red Triassic Mercia Mudstone below. The unconformity represents a gap in the geological record of about 120 million years.

Fossils

The last introduction to the basics of geology is the importance of fossils (named after the Latin *fodere*, 'to dig up'). As well as being fascinating and attractive objects in their own right, they are of great usefulness in geology, as a means for subdividing geological time, and also in revealing something about the environments in which rocks were deposited. In this last case use of the principle of uniformitarianism (see page 22) assumes that organisms found in ancient rocks lived in conditions similar to those of comparable types living in modern environments. For example, the reef-forming corals found in Carboniferous limestones 350 million years old lived in shallow warm tropical seas comparable to those in which the Australian Barrier Reef is forming today.

William Smith (1769-1839)

The 'Father of English Geology', William "Strata" Smith (1769-1839), working as a canal surveyor in England in the 18th Century, first realised that rocks of different ages contained distinctive and characteristic assemblages of fossils. He could identify and correlate rocks of similar ages in widely separated parts of the country by the fossils they contained. For example, rocks of Cambrian age are characterised particularly by trilobites, rocks of Jurassic age by ammonites etc. This fossil succession reflects the sequence of organic evolution – the natural appearance and disappearance of species through time.

Geologists use fossils to help them identify rocks of similar age. Rocks which are hundreds of miles apart can be recognised as being of the same age due to the fossils they contain. Ammonites in the Jurassic rocks are especially good for the purpose as many evolved and changed rapidly (in geological terms), producing different species with distinctive shell forms.

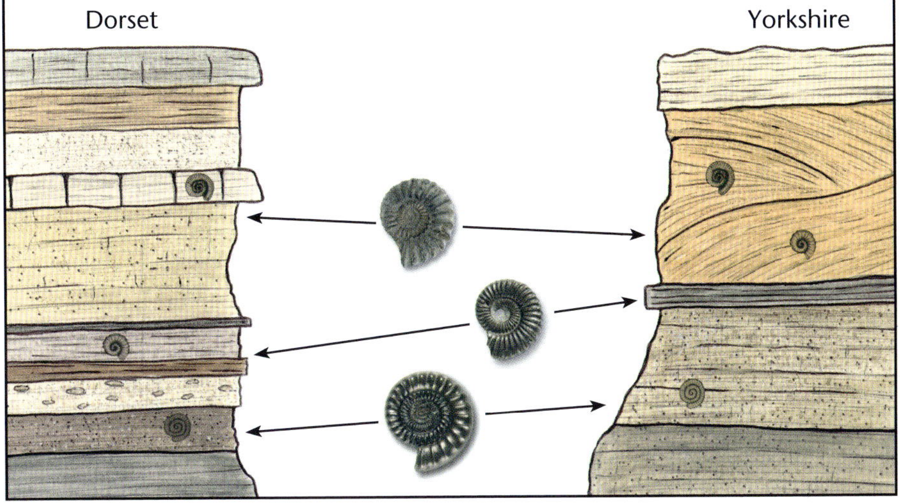

Fossils are of tremendous importance in correlation - that is, the matching-up of rock sequences between separate areas. Generally speaking, rocks containing the same fossil assemblages are similar in age. Another vital finding (called the principle of faunal succession) is that groups of organisms have succeeded each other during geological time in a definite order and each geological time period can be recognised by the fossils found in rocks of that age. Finally, we find that fossils in younger rock layers are closer to modern forms than fossils from older layers – a consequence of evolution.

Fossil echinoids (sea urchins) are useful for correlation within the Chalk

Since they were formed mainly on land, fossils are very rare in the red rocks. However, fossil reptiles, amphibians and fish are found in the Otter Sandstone and suggest an early Middle Triassic age (about 245 million years old). Around Exeter, fossil pollen in rocks below the Aylesbeare Mudstone indicates that they are no older than 270 million years (the early or middle part of the Middle Permian). The lack of useful fossils in strata between these two datable levels, including the Aylesbeare Mudstone and the Budleigh Salterton Pebble Beds, makes it difficult to decide on the position of the boundary between the Permian and Triassic. Traditionally it has been taken at the base of the Budleigh Salterton Pebble Beds, but another view is that the unconformable base of the Aylesbeare Mudstone represents the basal Triassic unconformity. However, studies of the magnetic properties (`palaeomagnetism') of the rocks by Mark Hounslow of Lancaster University suggest that the Aylesbeare Mudstone is Permian, with the boundary between the Upper and Middle Permian at a level around the base of the Littleham Mudstone (the upper half of the Aylesbeare Mudstone). In this view, the Permian-Triassic boundary is at the base of the Budleigh Salterton Pebble Beds, with a good part of the Upper Permian and most of the Lower Triassic missing at the boundary.

The situation is quite different in the Jurassic and Cretaceous rocks, which, being laid down in the sea, contain abundant remains of marine organisms including bivalves, brachiopods, sea urchins and ammonites.

Skull of the land reptile *Rhynchosaurus* found in the Otter Sandstone.

Specimen of a fossil rhynchosaur found at Ladram Bay in 1990.

The geological history of the Red Coast

This section examines in more detail the formations present along the Red Coast and the fascinating story that they tell us.

About 300 million years ago, South-west England was affected by earth movements called the Variscan Orogeny. Tremendous forces, driven by the movement of the 'plates' of rock that make up the earth's crust, caused a chain of highlands to rise over the region. Beginning in the Permian period and continuing into the succeeding Triassic period, the uplands began to be worn down, and vast amounts of gravel, sand and clay were washed down into the surrounding valleys and plains. During the Permian and Triassic the climate was semi-desert and most of the rocks are 'rusted' to the familiar red colours which characterise the soils and rocks of large parts of Devon.

Permian alluvial fans and muddy lakes – the Exe Breccia and the Aylesbeare Mudstone (about 265 to 255 million years ago)

The oldest rocks are the Exe Breccias around Exmouth which are Permian in age. They are a type of ancient gravel called breccia, laid down on great alluvial fans spreading out from the uplands that then existed to the west. They consist mostly of sharp-edged sandstone fragments, and lack of wear suggests that they were laid down rapidly.

The breccias are overlain by the Permian Aylesbeare Mudstone, about 400m thick, visible in the cliffs between Exmouth and Budleigh Salterton. They are mainly red mudstones with some thicker sandstones. These muds and sands were probably laid down in a type of inland lake that periodically dried out (a playa lake). It used to be thought that the sandstones were laid down by rivers, but this now seems less likely in view of the lack of channels and improbability that such a thickness (400m) of mudstone would be deposited on a floodplain without any channels. Instead, it now seems more likely that the sandstones were laid down by wind, representing the movement of sand dunes across the dry lake surface.

A Triassic gravel river – the Budleigh Salterton Pebble Beds (about 250 million years ago)

The fascinating Budleigh Salterton Pebble Beds are beautifully displayed in the cliffs west of Budleigh Salterton. They form a layer, up to 30m thick, of ancient gravels with some cross-bedded sand. The gravels are made up almost entirely of distinctive well-rounded pebbles of a very hard type of sandstone called quartzite. These gravels mark a sudden change in the environment. Until then, in Aylesbeare Mudstone times, the area was a quiet area of muddy inland lakes. Suddenly huge rivers flowing from the south swept in a great torrent of gravel. The fast-flowing rivers formed wide, gravel-choked, masses of interlacing braided banks and bars with water channels between. The size of the pebbles and cobbles, up to 18cm across, shows the power of the rivers that carried them. Interestingly, the pebbles and cobbles get smaller as the Pebble Beds are traced northwards, supporting the idea that the rivers flowed from the south so that the coarser material would be deposited first. At this time the area that we now call the English Channel was all land, across which the rivers flowed.

The quartzite pebbles are of particular interest because they contain fossils – including brachiopods, bivalves and trilobites. But these animals did not live at the same time as the Pebble Beds were forming. The sources of the pebbles and their fossils have been a subject of great interest to geologists over the years. Research shows that the fossils are of Ordovician and Devonian ages and can be matched with fossils from rocks in France (Brittany and Normandy) and south Cornwall. This is further evidence that the sources of the great gravel rivers were in the south.

The Pebble Beds rivers seem to have stopped flowing almost as suddenly as they started. We can tell what happened next by examining a thin layer of red clay that occurs right at the top of the Pebble Beds and which can be seen in the cliff west of Budleigh Salterton. This layer, only 0.3m thick, is a fossil soil that formed on top of the dried-up river gravels. Capping this soil are pebbles of quartzite with curious facets that were produced by sand blasting in a desert. These are called ventifacts, and the bed with them can be traced at least as far north as Uffculme. We can now visualise much of East Devon at this time as a wind-blown stony desert.

An example of a ventifact, about 8cm long.

A Triassic sand river – the Otter Sandstone (about 247 to 240 million years ago)

Following the stony desert conditions after the Pebble Beds, there was a gap in deposition of uncertain length. Above come the sandstones of the Otter Sandstone, which are up to about 210m thick. They are mainly reddish brown sandstones made up of fine to medium-grained particles, with some thin layers of hard conglomerate and a few discontinuous layers of red mudstone. Cross-bedding is quite common in the sandstones.

At Budleigh Salterton, the first 14m of deposits immediately overlying the Pebble Beds are wind-blown sands with distinctive cavernous weathering (see photos on pages 64 and 65). These sand dunes were only local, however, and were soon covered by sands laid down by great

Cross-bedding is clearly visible in this sandstone.

rivers that began to flow into the area. They flowed from south to north, like the earlier Pebble Beds rivers. The climate was arid to semi-arid, and the rate of flow of the rivers was highly variable. The sandy rivers flowed in shallow braided channels, and in the present-day cliffs of our area we can recognise 'stacks' of these channels. A typical channel, where fully developed, begins with a 'lag' conglomerate at the base, resting on an erosional surface cut into the deposits beneath, and passes upwards into cross-bedded sandstone and then into mudstone.

Vegetation, consisting of plants with deep taproots adapted to semi-arid or arid conditions, probably grew profusely on sandbars and in abandoned river channels. Evidence for this is provided by the presence, especially in the lower part of the Otter Sandstone, of vertical tubular calcareous concretions which might have formed around the roots of such plants (see photo, page 66). There are also horizontal or inclined calcareous sheets that may have been deposited from carbonate-rich groundwater. From fossils found in the Otter Sandstone we know that there was quite a variety of animals living on the river plain or in the fresh water, and the remains of arthropods, fish, amphibians and reptiles have been discovered. In fact the Otter Sandstone is of international importance for the fossils of Mid Triassic age that it has yielded, although they are mostly found only as fragments in the conglomerate layers at the base of channels.

Especially notable are the strange rhynchosaurs, a type of land reptile which we can envisage lumbering around the sandy river plain. These reptiles, about 2m long, predated the dinosaurs. They were herbivores with horny beaks at the front of the mouth and many rows of blunt teeth at the back. They probably ate horsetails and other plants such as seed ferns and cycads. The unusual curved snout of the animal, clearly visible in the photo of a fossil rhynchosaur skull on page 30 gives rise to the name rhynchosaur which means 'snouted reptile'. The origin of rhynchosaurs is uncertain, and they disappeared without trace at the end of the Triassic. An excellent specimen (*Rhynchosaurus spenceri*) was found at Ladram Bay in 1990, in place on a sandstone ledge surrounded by shingle, and is shown on page 31. The specimen is now in the Royal Albert Memorial Museum, Exeter. A fascinating reconstruction of Mid Triassic life during Otter Sandstone times is shown opposite.

Triassic lakes – the Mercia Mudstone (about 240 to 209 million years ago)

Towards the top of the Otter Sandstone there is a transitional interval of 20-30m between the main sandstone-rich part of the Otter Sandstone and the overlying Mercia Mudstone. This interval represents interplay between the rivers of the Otter Sandstone and the lakes of the Mercia Mudstone. However, the final sand influx is quite clear and the last sandstone which marks the top of the Otter Sandstone can be clearly seen (for example, in the cliffs below High Peak, Sidmouth and in the cliffs just east of the mouth of the River Sid). The scene was one of much quieter deposition, and much of the Mercia Mudstone formed in a large playa lake. The latter is a shallow lake in a semi-arid or arid region of a closed desert basin which dries up at times.

Reconstruction of Mid Triassic life during Otter Sandstone times. Two rhynchosaurs are pictured in the middle distance, and behind them is a pair of rauisuchians. The scorpion on the nearest rock looks towards a pair of procolophonids on the rocks. Opposite them is a large temnospondyl amphibian. The plants include horsetails at the waterside and a type of conifer.

The Mercia Mudstone, up to about 440m thick, is almost entirely reddish brown mudstone. Within it are two subdivisions, one in the middle and one at the top, which consist largely of green mudstone with common beds of dolomitic mudstone and less common beds of purple and/or dark grey mudstone. The middle unit, the Dunscombe Mudstone, is a distinctive layer of about 35m of green and purple mudstones. It may have formed in a series of shallow freshwater lakes crossed at times by broad shallow channels bringing in sand. Conditions were probably wetter and cooler than those in the red beds above and below. The Dunscombe Mudstone occurs in the cliffs below Higher and Lower Dunscombe cliffs, and east of Weston Mouth comes down to beach level below Coxe's Cliff. This is the only part of the Mercia Mudstone that has yielded fossils visible without the use of a microscope, such as specimens of the branchiopod crustacean *Euestheria* and various trace fossils.

The highest unit, the Blue Anchor Formation, 29m thick, is seen below Haven Cliff, east of Seaton. It records a gradual change upwards in environment from deposition on land to deposition in a sea; it is overlain by the black mudstones and limestones of the Penarth Group, mainly laid down in a sea.

In the remaining red part of the Mercia Mudstone careful study and detailed measurement shows that several subdivisions can be recognised, based on characteristics such as the presence of gypsum, minor sandstones etc.. For those interested in more detail, there is a full account by Dr Ramues Gallois (see Further Reading section, on page 124). The Mercia Mudstone has in places yielded microscopic miospores and other remains of land plant origin. These are useful for dating the parts of the rock sequence that they occur in.

When mineral-rich waters evaporated in the hot desert sun, evaporite minerals such as gypsum formed. It occurs scattered throughout the Mercia Mudstone, but is commoner at some levels. At Red Rock [SY 191 880] near Branscombe, a 10m cliff shows gypsum veins interwoven with red mudstone. Hutchinson, in his History of Sidmouth, recorded the remains of ancient salt crystals forming cubic impressions in mudstone near Salcombe Mouth, and also in the cliff beneath Windgate, west of Sidmouth.

At a few places, current ripple marks formed by the movement of water can be seen.

Red Rock [SY 191 880] near Branscombe. A 10m cliff shows gypsum veins interwoven with red mudstone.

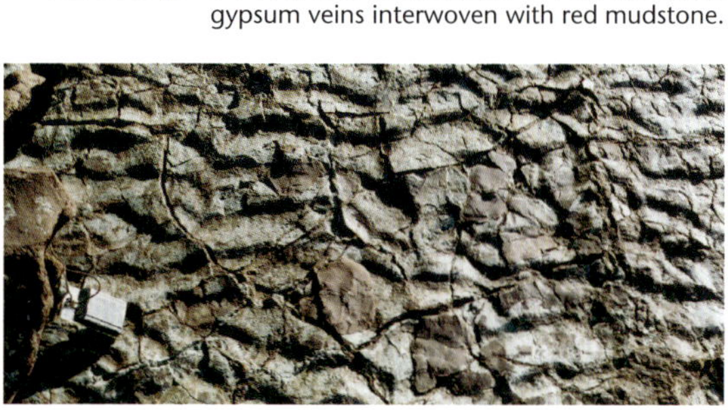

Ripple marks in the Mercia Mudstone, in a fallen block below Peak Hill.

The change to marine conditions - the Penarth Group (about 209 to 201 million years ago)

Taken together, the Blue Anchor Formation and Penarth Group record a change from the largely continental environment of the red Mercia Mudstone, to gradually more marine conditions. The Westbury Formation is the lowermost subdivision of the Penarth Group and consists of black mudstones laid down in a brackish to marine environment. The overlying limestones and mudstones of the Lilstock Formation were laid down on the sea-bed.

The Jurassic sea (about 201 to 145 million years ago)

The Penarth Group is succeeded by Jurassic rocks, exemplified by the classical exposures of the Blue Lias and Charmouth Mudstone at Lyme Regis and at Pinhay Bay at the eastern limit of our area. By the time the Blue Lias was laid down, a shallow sea extended over much of Britain. The sea was full of marine life, including ammonites and marine reptiles such as ichthyosaurs and plesiosaurs. A small selection of Jurassic fossils is shown here.

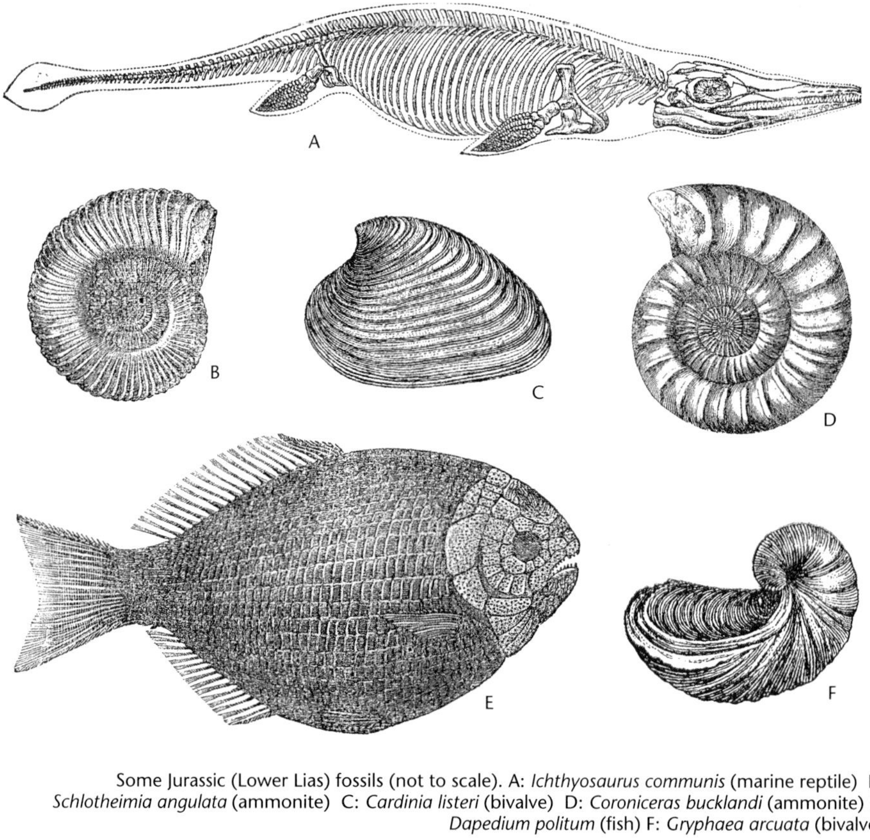

Some Jurassic (Lower Lias) fossils (not to scale). A: *Ichthyosaurus communis* (marine reptile) B: *Schlotheimia angulata* (ammonite) C: *Cardinia listeri* (bivalve) D: *Coroniceras bucklandi* (ammonite) E: *Dapedium politum* (fish) F: *Gryphaea arcuata* (bivalve)

Cracks in the rocks

Faults are cracks in the earth's crust along which rocks have moved relative to each other. The illustration here shows how a typical 'normal' fault displaces the geological layers in a block of rocks. Several large faults as well as some minor ones affect the rocks. The bigger faults are shown on the geological fold-out map at the back of this book. The most important ones are: the fault at Jacob's Ladder, Sidmouth, that brings the Mercia Mudstone on the west against the Otter Sandstone on the east, and the fault at Seaton Hole that brings Upper Greensand on the west against Mercia Mudstone on the east. Other noteworthy smaller faults are present at the following places: around Orcombe Point, in the cliffs of Sandy Bay, and at Straight Point, several north to south trending faults affect the Aylesbeare Mudstone; small faults affect the Otter Sandstone just east of the mouth of the River Otter, at Otterton Point and Danger Point; in the cliffs at Sandy Cove, just east of Ladram Bay, where Otter Sandstone on the north is faulted against Mercia Mudstone on the

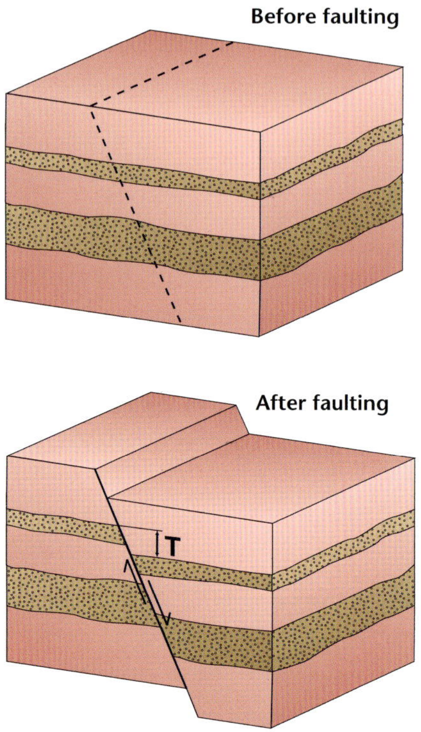

Block diagram showing the effect of a 'normal' fault on layered rocks. The 'throw' of the fault, T, is the vertical change in level of the displaced layers.

south; just east of the mouth of the River Sid. Faults also extend along the Salcombe Valley and the Weston Mouth Valley. East of Seaton, north-south - trending faults are seen in the Rousdon and Pinhay areas.

Tilting of the rocks

As well as fracturing rocks, earth movements can also tilt or bend them. The Permian and Triassic rocks show an easterly tilt, of about 5 degrees on average. When did the tilt happen? This can be decided by finding the youngest beds that are involved in the tilt, which prove to be the Wealden (Lower Cretaceous) rocks of the Swanage area. The tilt must have taken place after the date of those rocks – probably about 120 million years ago.

Between Branscombe and Beer, the Cretaceous rocks (Upper Greensand and Chalk) have also been bent by earth movements into a shallow bowl shape called a syncline. The rocks on either side of the syncline dip inwards towards its centre – for example, the Cretaceous rocks at Beer Head dip to the east while those at Whitecliff dip westwards. This structure, called the Beer Syncline, is made clearer on the cross-section at the back of this book, and can be seen in the photo on page 42.

East of the mouth of the River Axe, the Mercia Mudstone has been bent into a shallow convex-upward fold called an anticline (shown in the photo on page 97).

The Gault and Upper Greensand sea (about 113 to 100 million years ago)

About 113 million years ago, the Site was invaded by a shallow sea (a process called a marine transgression), which extended over much of South-west England, although we are not sure whether Dartmoor stood up as an island. The first deposits in the shallow sea were muds, now represented by the thin Gault. The Gault was laid down

A charming 1823 drawing of the East Devon coast by William Buckland, showing well the main geological subdivisions: Triassic rocks - pink; Upper Greensand - green; and Chalk - white.

only patchily, and nowhere in our area can you see it in place. Its presence in the Beer area is inferred through the presence of the large landslip at The Hooken, which might have moved on a thin layer of slippery Gault.

Later, sands were laid down on the seabed, and these gave rise to the formation which we call the Upper Greensand. The Cretaceous sediments gradually overstep the bevelled edges of successively older formations westwards until, in the area near Dartmoor, they rest on Devonian rocks. This gives a classic example of an angular unconformity. The unconformity is made even more striking in the area between Sidmouth and Branscombe by the colour contrast between the red rocks and the overlying yellowish brown and white Cretaceous rocks. The charming 1823 drawing on pages 38-39 of the East Devon coast by William Buckland shows well the main subdivisions.

View from Sidmouth looking eastwards, showing pale Cretaceous rocks in the upper part of the cliffs unconformably overlying red Triassic rocks in the lower cliffs.

The Upper Greensand is 45 to 55m thick and is divided into three parts, in ascending order: the Foxmould (20-25m thick), Whitecliff Chert (up to 32m thick), and Bindon Sandstone (up to 8m thick). The Foxmould is mainly calcareous sandstones and calcarenites between Salcombe Hill and Beer, but is mainly sand at High Peak and at localities west of the Sid Valley. It weathers to a 'foxy' yellowish brown colour (hence the name), but where fresh is greenish grey. The Whitecliff Chert consists of sandstones and calcarenites with many layers of nodular and tabular chert. The Bindon Sandstone also contains some chert in its lower part, but this disappears west of Beer Head.

The sands and sandstones consist mainly of quartz grains. A significant but small component, seen as small dark green specks, is the mineral glauconite, which generally forms only in marine conditions. The photograph here shows grains of dark green glauconite in a fallen block of fossiliferous Foxmould. The sands and sandstones, especially of the Foxmould, have been extensively disturbed by burrowing animals, a process called bioturbation, and this has usually destroyed any original

Grains of dark green glauconite in a fallen block of Foxmould with beautifully preserved marine shells.

structures such as cross-bedding. Another feature of the Upper Greensand, not seen in the Otter Sandstone, is the presence of fossil shell material at many levels.

The Upper Greensand contains many different types of fossil marine bivalves, with relatively common brachiopods, gastropods and echinoids. Unlike the Jurassic rocks, ammonites are scarce. Fossils can be seen in fallen blocks on the beach, particularly at Whitecliff and along the beach between Branscombe and Beer Head. Inland, many specimens were collected from the Blackdown Hills, and the Royal Albert Memorial Museum in Exeter has a fine collection. The photo on the previous page shows beautifully preserved marine shells in a fallen block of Foxmould, on the beach below Higher Dunscombe Cliff. The structures in the Upper Greensand, combined with the environmental evidence provided by the fossils, suggest that it was laid down in a shallow current-agitated sea. The cherts in the Upper Greensand show traces of bedding, trace fossils and other fossils suggesting that they represent replacement of existing sediments such as sandy calcarenites, by silica.

The Chalk sea (about 100 to 86 million years ago)

Following Upper Greensand times, the area remained under the sea, but a quite different type of material was deposited on the sea floor – the Chalk. Sand grains are almost completely absent from Chalk, and it consists largely of the microscopic remains of tiny organisms called algal coccoliths. These slowly drifted down, like snow, to the sea-floor in incalculable numbers to accumulate into a pale mud which later hardened to form the Chalk that can be seen today, most spectacularly around Beer Head.

Electron microscope photograph showing coccoliths. Much of the Chalk is made up of minute organisms similar to these.

Conspicuous bands of flints occur within the Chalk at certain levels. They consist of silica (silicon oxide) which replaced a pure, very fine-grained, carbonate rock. Hence, unlike cherts, flints rarely show bedding traces. Almost all flints start as replacement of trace fossils such as burrow fills; some then grow to irregular mats or concretions. Sometimes fossils such as sea urchins (echinoids) are replaced by flint.

Flint bands in Chalk near Beer Head.

At the base of the Chalk, and resting directly on the Upper Greensand, is

A view looking across Beer Beach. The cliff in the foreground shows Upper Greensand overlain by Beer Head Limestone and Middle Chalk. In the background the cliffs show Middle Chalk overlain by the Upper Chalk of Annis' Knob. This photo also illustrates the structure of the Beer Syncline, in which the layers in the nearest cliff dip away from us (eastwards) but towards us (westwards) in the farthest cliffs.

a bed of limestone called the Beer Head Limestone, varying between 0.3 and 12.6m thick. It is beautifully displayed on the west side of Beer Beach, where it forms a thin (0.6m) layer resting on a hardened surface of Upper Greensand. There was probably a break in sedimentation between the two.

Detailed view of the Beer Head Limestone on the west side of Beer Beach. It rests sharply on the Upper Greensand, and is overlain by Middle Chalk.

The Beer Head Limestone contains many fossils, including ammonites, brachiopods, echinoids, gastropods and sponges. The ammonite assemblage is very diverse and includes forms not known elsewhere in Britain. The Beer Head Limestone is the equivalent of the Lower Chalk elsewhere in England. It can be divided into three parts, A, B and C, separated by mineralised hardground surfaces. Division A is a hard white shelly limestone. The lower part contains quartz grains and pebbles and also quite common examples of the coral-like fossil bryozoan *Ceriopora ramulosa*. Division B is a hard white sandy glauconitic limestone, `welded' on to the underlying rock of Division A. Division C is very locally developed in places where the Beer Head Limestone is thicker, as at Beer Head; it consists of very hard nodular sandy and glauconitic limestone. Where Division C is absent, the Middle Chalk (*Inoceramus labiatus* Zone) rests directly on an erosion surface at the top of Division B. The Beer Head Limestone was deposited in a shallow sea.

Overlying the Beer Head Limestone is much more typical Chalk - white, very fine-grained limestones. This part of the Chalk was formerly divided into the traditional subdivisions of Middle and Upper Chalk. Most of the cliff at Beer is Middle Chalk, but the upper cliffs, for example, Annis' Knob, are in Upper Chalk. There is a prominent band of black flints running through Annis' Knob. The traditional chalk subdivision of Lower, Middle and Upper Chalk has now been replaced by a new system of names. For example, the Middle Chalk is now divided into two formations: the Holywell Nodular Chalk overlain by the New Pit Chalk, and the Upper Chalk is now termed the Lewes Nodular Chalk, overlain by Seaford Chalk. These names are similar to Chalk sequences elsewhere in southern England and across the Paris Basin, showing how uniform the conditions of climate and water depth were at that period of earth history.

Fossils in the Upper Greensand and Chalk include broken bivalve shells, brachiopods and fossil sea urchins (echinoids). Some characteristic Upper Greensand and Chalk fossils are shown below.

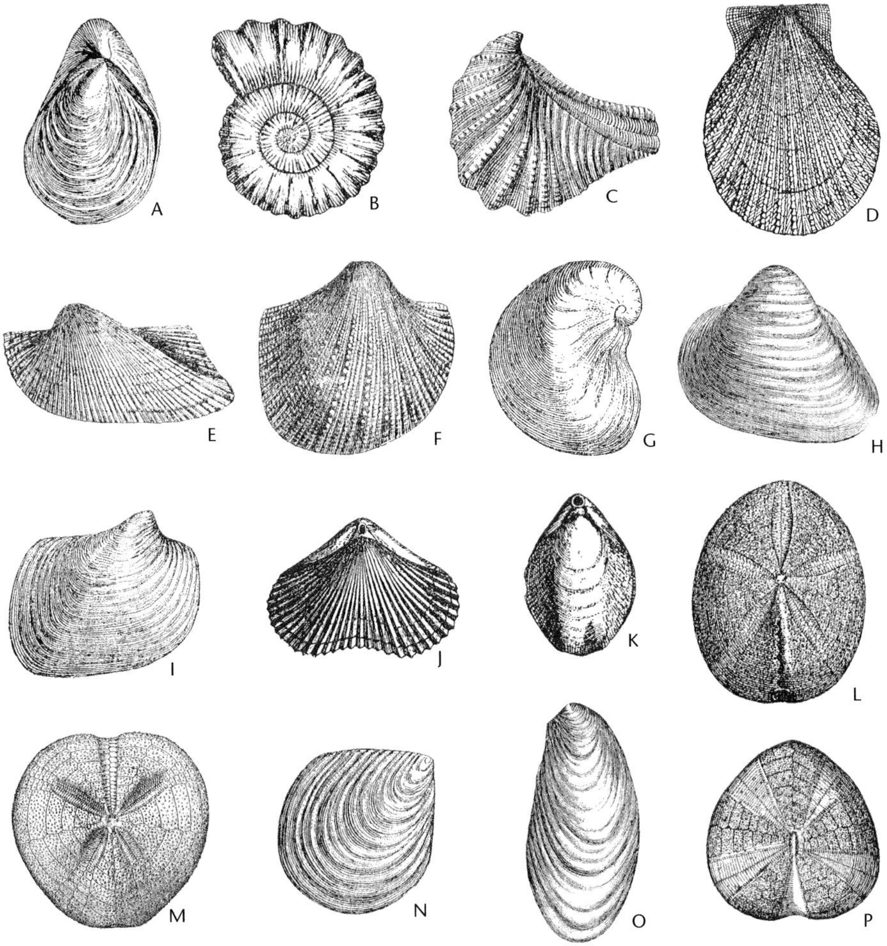

Some of a wide variety of Upper Greensand and Chalk fossils (not to scale). **Upper Greensand** A: *Actinoceramus concentricus* (bivalve). B: *Hysteroceras varicosum* (ammonite). C: *Pterotrigonia aliformis* (bivalve). D: *Spondylus dutempleanus* (bivalve). E: *Nannonavis carinata* (bivalve). F: *Cardium* (*Granocardium*) *proboscideum* (bivalve). G: *'Exogyra conica'* (bivalve). H: *Cucullaea glabra* (bivalve). I: *Panopaea mandibula* (bivalve). J: *Rhynchonella dimidiata* (brachiopod). K: *Ovatathyris ovata* (brachiopod). L: *Catopygus columbarius* (echinoid). **Chalk** M: *Micraster cortestudinarium* (echinoid). N: *'Inoceramus latus'* (bivalve). O: *Mytiloides mytiloides* (bivalve). P: *Crassiholaster subglobosus* (echinoid).

The Clay-with-flints

Close up of typical Clay-with-flints, showing angular unworn flints in a matrix of reddish brown clay.

The Clay-with-flints caps the flat-topped Upper Greensand and Chalk hills between Salcombe Hill and Beer and between Seaton and Lyme Regis. It consists of brown clay with unworn flints. It normally forms a layer up to about 5m thick, but in places it fills deep solution pipes that descend into the Chalk or Upper Greensand to considerable depths, in places as much as 30m. There has been much discussion about how the Clay-with-flints formed. The most likely sequence of events seems to be that, first, the Chalk was uplifted to form dry land, followed by erosion to form a flat plain cut across the Upper Greensand and Chalk. Then, in Tertiary times, rivers flowed across the plain to deposit sheets of gravel, sand and clay. There was then a period during which the Chalk and Greensand beneath the gravels were dissolved and mixed with the insoluble residues of the Chalk and Greensand.

A particularly interesting feature of the Chalk, seen in the upper cliffs between Beer Head and Whitecliff, is a distinctive 15m- to 20m- thick layer of Chalk which has been partly dissolved. The major Chalk solution features at Beer were first recognised by Dr Ramues Gallois, and he has provided the description of this fascinating feature of the geology.

The solution layer consists of angular to rounded pieces of fairly unaltered chalk, up to boulder size, in a matrix of fine-grained chalk debris. Large blocks of intact unweathered chalk up to tens of metres thick can be seen 'floating' in the partly dissolved layer. The well-known feature of Annis' Knob is an example of such a chunk of 'floating' chalk. The lower limit of the partly dissolved layer is sharply limited by a thin bed (only 5-10mm thick) of clayey chalk (the 'Dowlands Marl') that has acted as a barrier to the downward flow of the dissolving waters and which extends for over

Pipes of Clay-with-flints (CF) near Coxe's Cliff. The pipes descend into the white Chalk (C). Beneath the Chalk is Upper Greensand with the Bindon Sandstone (BS) having a smooth appearance on rougher Whitecliff Chert (WC) beneath.

A view of the cliffs of Chalk at Beer, showing the solution layer at the top of cliff. Upper Greensand (UGS) and Beer Head Limestone (BHL) occur at the base of the cliff.

A closer view of the solution layer in the upper part of the Chalk cliffs, above Pound's Pool Beach, near Beer. In the far distance are the red Mercia Mudstone cliffs of Seaton.

2km in the Beer area. An important practical point arising from the recognition of the partly dissolved Chalk layer is that it is relatively weak and so is more prone to landslips than the unweathered chalk, especially after heavy rain. Collapses can happen very quickly and some have been on a large scale. For example, in March 1933 a huge slip covered the west side of Beer Beach and there have been many slips since then. The most recent case of a slip associated with the solution layer happened in 2001 below Annis' Knob and led to the closure of the cliff path.

The two photographs above show the cliffs at Beer and the solution layer described above. The solution layer in the closer view is particularly well seen, and the tilt of bedding of areas within the layer, such as that in the top left hand corner of the photo, shows how Chalk blocks have foundered into it.

An interesting rock type, found widely around Sidmouth, is silcrete (see photos on page 26). It is made up of angular flints set in a compact pink to grey fine-grained matrix of silica-cemented silt-sized quartz and flint grains. The rare presence of fossilised plant root structures shows that silcrete formed on or near a land surface. Its origin is complex, but it appears to have formed in dry periods when silica was deposited as the soil profile dried out.

West of Sidmouth silcrete occurs mainly on the Bulverton Hill - Mutters Moor - Peak Hill plateau, and east of Sidmouth between Salcombe Hill and Trow. It forms very hard blocks, sometimes large, lying on the surface of the plateau. Many have been removed by farmers and also, during Victorian times, many large blocks were carted off to build rockeries. An example is the grotto at Knowle, now part of the garden at Knowle Grange in Sidmouth. The silcrete stones making up a prehistoric stone circle ('The Seven Stones') on Mutter's Moor were taken away in 1830 by Lord Rolle to make a rockery and shell grotto at Bicton Park.

The last two million years – the 'Ice Age'.

The last two million years was a period known as the Quaternary, sometimes called the Ice Age, and was a time of tremendous and sometimes rapid climate change. The Quaternary is currently taken to represent the last 1.8 million years, but may in the future be amended to represent the last 2.6 million years. Temperatures oscillated between cold (glacial) periods and more temperate (interglacial) periods. During the cold periods in the northern hemisphere, great ice sheets and valley glaciers advanced from the north and then retreated. Between these periods of severe cold, the climate became at times even warmer then today. For example, in our own area, excavations for the Honiton bypass road yielded deposits with the remains of hippopotamus, elephant, giant ox and red deer, indicating an interglacial period about 130,000 years ago.

Handaxes from Broom, possibly at least 280,000 years old.

A major advance in understanding Quaternary climate change came from looking at cores from boreholes drilled into ocean floor sediments and also through the polar ice caps. Measurements of the ratio of two stable isotopes of oxygen in the cores showed dramatic fluctuations related to global temperature, indicating numerous alternations between cold and warm periods. This research suggests that during the last 2.6 million years there may have been as many as 50 glacial stages and a similar number of temperate stages - many more than previously suspected from evidence on land. The stages are called Oxygen Isotope Stages (abbreviated to OIS). A curve showing the cold and warm stages over the last 650,000 years is shown opposite.

Oxygen Isotope Stages of the Quaternary

and corresponding British glacial and interglacial events

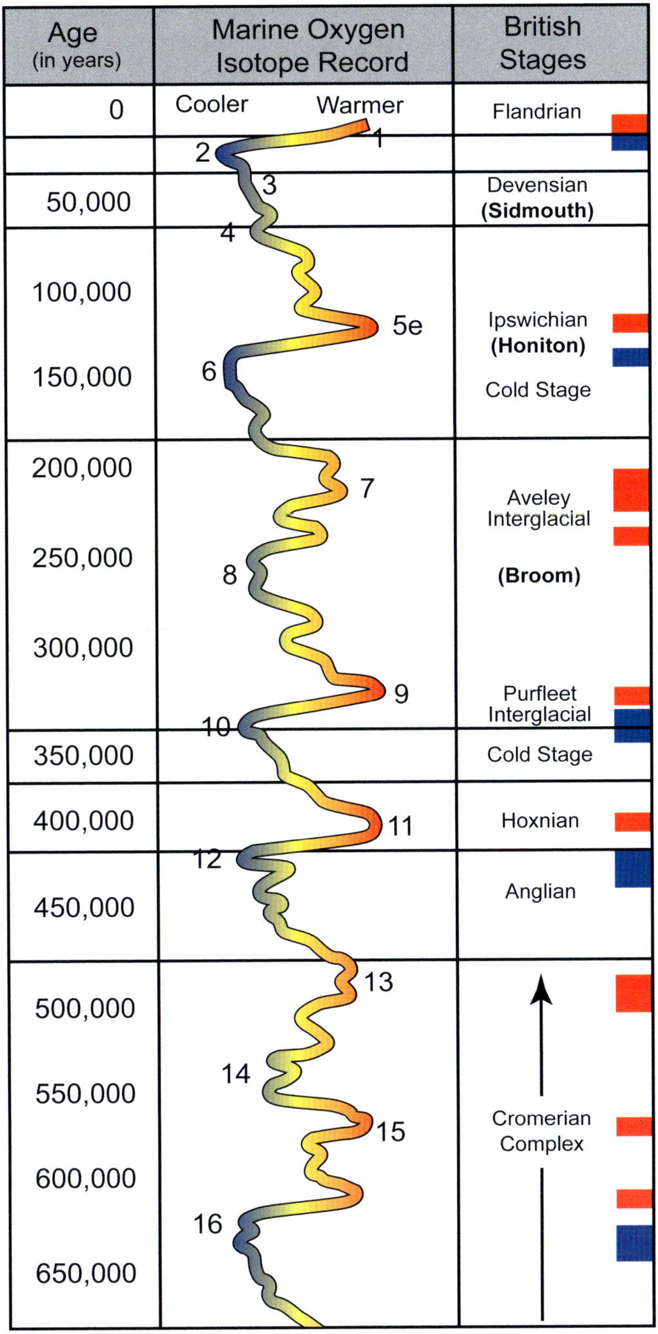

The cold periods are evenly numbered and the warm periods are given odd numbers. Hence the present warm interglacial period is OIS 1 while the last glacial period is OIS 2. Relating the marine record to that on land is difficult because of the fragmentary nature of the deposits on land - the earlier deposits were often destroyed as the next ice sheet advanced southwards. However, some progress has been made in linking the onshore evidence with the oxygen isotope evidence, especially over the past 650,000 years or so. The onshore British glacial and interglacial stages for this period and their possible identification with the numbered OIS stages are shown in the illustration on page 47.

Mammoth tooth in Sidmouth Museum.

Within East Devon, river terrace gravels at Broom in the Axe valley have yielded many examples of the famous handaxes (pictured on page 46) fashioned from Greensand chert. These may date from towards the end of the cold stage OIS 8, about 280,000 years ago. As noted above, we have evidence of Ipswichian interglacial deposits (OIS 5e; about 130,000 years old) from the Honiton bypass. Clays with Mammoth teeth found under Sidmouth beach in 1873 probably indicate the middle of cold stage OIS 3, identified with the most recent (Devensian) glaciation. A typical Mammoth tooth is pictured here.

About 10,000 years ago, the last ice sheet melted and the climate became more temperate. During this last glaciation, the World Heritage Site lay just south of the ice sheets which actually pressed right up against the north coast of the South-west England peninsula. The nearness of the ice sheets had a tremendous effect on the deposits and landscape of the area. During cold periods, the area had a periglacial climate rather like parts of the Canadian Arctic or Siberia today, and much of the ground was deeply frozen (permafrost).

A wide variety of deposits that formed during the Quaternary can be found, and some, such as beach deposits, are still forming today. The following types can be recognised:

Deposits formed by the sea or in the intertidal zone.
These include the modern beaches and shingle ridges along the coastline, and also the saltmarsh deposits (some of which have been reclaimed for agriculture) of the River Otter and River Axe valleys.

Deposits formed by rivers.
These include river terrace gravels, which are the dissected remains of former floodplain deposits of rivers, now left at levels higher than the modern alluvium (the name given to the deposits being laid down on modern river floodplains – mainly clay, silt, sand, gravel and peat). The process of river terrace formation is illustrated opposite.

The formation of river terraces

Early in the evolution of a valley, a river flows over a floodplain.

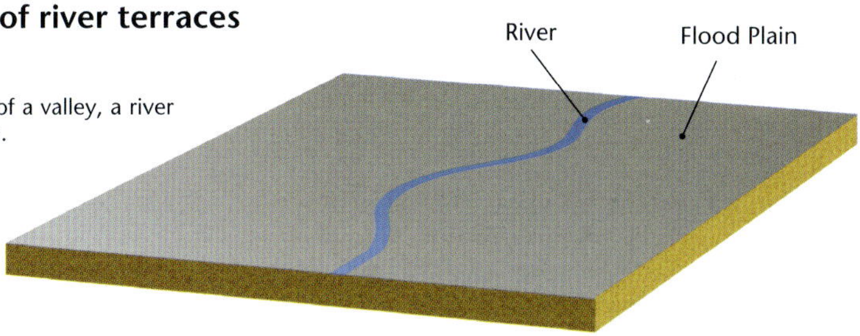

Rejuvenation of the river, perhaps by a change in sea level, causes it to cut down into the former floodplain, leaving its remnants as terraces on both sides of the valley.

Renewed down-cutting leaves a second pair of terraces, at a lower level than the first. This process may be repeated several times, and in the Otter Valley up to 10 levels have been recognised, the highest being the oldest.

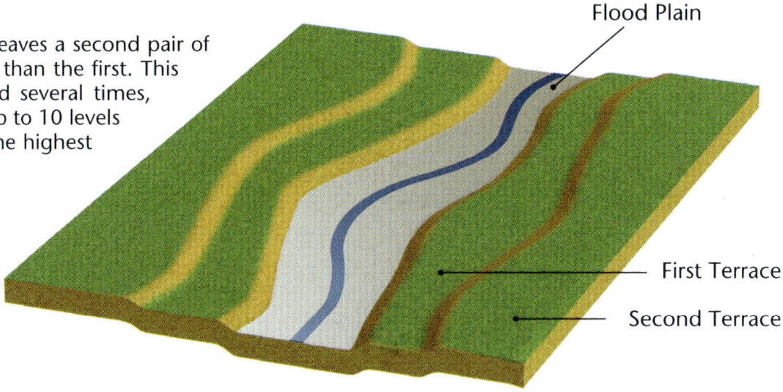

Later erosion isolates the terrace deposits to form a chain of isolated remnants along the valley sides. This is the current situation in the Otter Valley.

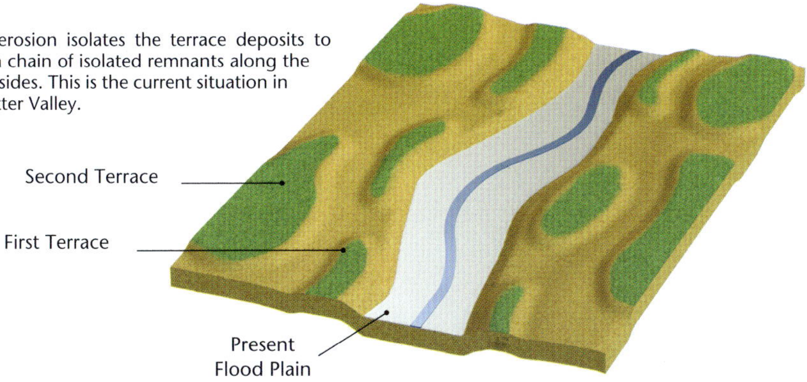

River terraces are present particularly along the course of the River Otter. They can be seen on the coast between Otterton Ledge and Ladram Bay where the sea has cut into the cliffs. The river gravels represent the deposits laid down by the River Otter in the last few hundred thousand years when it flowed at a higher level and had a course different from that today. The different levels are called river terraces, since they sometimes form a series of successively higher step-like features. There are up to ten levels of river gravels along the course of the River Otter, the highest being at Belbury Castle [SY 076 941, near West Hill, where the terrace is at a height of about 107m above sea level. The distribution of the river terraces is shown on the geological map at the back of this book. In the Axe Valley, the river terrace gravels at Broom are notable for containing large hand axes, possibly made by early man (*Homo heidelbergensis*) at least 280,000 years ago. The implements are mostly made out of Upper Greensand chert, and good examples can be seen in the Royal Albert Memorial Museum Exeter.

Deposits formed by mass movement processes - mainly head and landslips

The curious term 'Head' was first used by the famous geologist Henry De la Beche for rubbly slope deposits which occur along the coast of SW England. It is now used for deposits which probably formed during cold periods of the Ice Age. Then, alternate freezing and thawing of surface layers resulted in a porridge-like mass of waterlogged soil and other material, which moved down slopes. Head is one of the most widespread of the Quaternary deposits and consists of clay or sand with pieces of resistant rocks such as flints or cherts derived from the Chalk and Upper Greensand. It occurs just about everywhere, forming a kind of skin over most of the solid formations.

Typical Head overlying Otter Sandstone just east of the mouth of the River Otter.

Deposits formed by wind

An area of wind-blown sand has formed near the coast of Exmouth in an area called The Maer, and is shown on the geological map at the back of this book.

Landslips are found mainly along the coast, for example near Beer Head where the Hooken Landslip is a striking sight. The whole stretch of coast between Seaton and Lyme Regis is intensively landslipped and includes the site of the famous Bindon Landslip.

The Bindon Lanslip, from a contemporary print by Hullmandel 1839-40. The area is now heavily overgrown with trees.

The photo here shows a spectacular example of a dust cloud produced by a rockfall at Salcombe Hill Cliff on 25th June 2001, seen from Sidmouth. The rockfall was caught on camera by Pamela Harrison. Despite the dust cloud relatively few blocks fell to the beach and the cliff path was not affected.

During the greatest extent of the last ice sheet, about 18,000 years ago, vast amounts of water were locked up in the ice and sea level fell all over the world by at least 100m. When the most recent ice sheet started to melt, about 10,000 years ago, sea level began to rise, rapidly between 10,000 and about 5,000 years ago, and much less rapidly from then until the present. As the sea rose, woods around the coasts were drowned. Remnants of these 'submerged forests' are present all round the shores of South-west England. Possible examples have been recorded at Sidmouth.

Dust cloud from a rockfall at Salcombe Hill Cliff, June 2001.

The landscapes of the Red Coast

The Red Coast landscapes are notable for the close relationship that exists between the rocks and the shape of the land, shown in the diagram on page 54.

Away from the coast, of the Permian and Triassic formations, the Budleigh Salterton Pebble Beds have the most distinctive expression in the landscape. This arises because of the contrast with the underlying red mudstone, so that the Pebble Beds, which are more resistant to weathering, form a distinctive escarpment rising in places to about 30m above the mudstone vale below. It can be seen, for example, on the west side of Woodbury Common (car park at SY 031 872). Also, because of the general infertility of the Pebble Beds soils, they give rise to heathland so that there is a contrast in land use between the farmland of the mudstone vale and the Pebble Beds heathlands. Thus a chain of heathlands, including Woodbury Common, extends along the Pebble Beds outcrop.

Inland from the coast, the area is basically a flat-topped plateau rising to about 246m above sea level, formed of Upper Greensand and Chalk capped by Clay-with-flints. The Clay-with-flints forms heavy, very flinty soils, and although some of the plateau is used for farming, much is taken up by forest plantations (for example, on Bulverton Hill, west of Sidmouth). Where fields are ploughed, the flintiness of the soils is all too apparent! In its natural state the plateau forms heathland, but little is left; a surviving area can be found on Mutter's Moor [SY 105 881], west of Sidmouth.

The plateau was formerly continuous, but streams have cut down into it to form valleys such as the Sid Valley. The present-day plateau is by no means a continuous surface at the same elevation everywhere, but has been dropped down in places to lower levels by faults. For example, the East Hill Plateau north of Sidmouth is at a level of up to 246m above sea level, whereas the Salcombe Hill Plateau east of Sidmouth is at a maximum elevation of only 176m above sea level.

On the valley sides, the contrast in scenery between the Cretaceous rocks and the Triassic rocks is very clear. The Cretaceous rocks are more resistant than the underlying Triassic mudstone and so form steep slopes, up to 30m high, which are often wooded. There is a distinct break of slope at the base of the greensand, with less steep slopes on the red mudstone beneath. Rainwater that has drained down through the porous greensand is thrown out as springs where it meets the impermeable mudstone beneath. This spring line is known locally as the 'Moor Ditch'. There is also a change in land use, with the wooded greensand slopes contrasting with the farmland of the red mudstone below. The band of steep wooded greensand slopes bordering the valleys is one of the most characteristic features of the inland East Devon landscape.

The streams and their tributaries originate in springs at the base of the Upper Greensand, in very steep sided deeply cut little gorges which are locally called 'goyles'. The source of the River Sid is beside the road to Ottery St Mary, about 800m west of the Hare and Hounds pub, Gittisham Hill, and there is a small sign marking the spot. Just south of the road, the infant stream descends into a deep goyle. Another notable goyle is that at the source of the Roncombe Stream [SY 169 943] near Roncombe Gate.

Very flinty soils on the Clay-with-flints plateau. Sincombe Hills [SY 119 918].

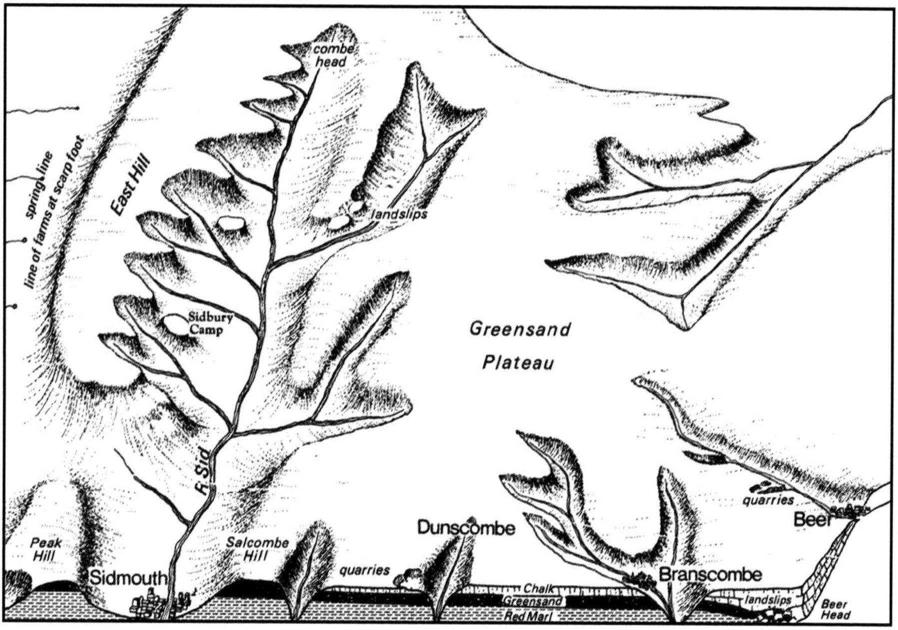

Block diagram of the area between Sidmouth and Beer Head showing the relationship between geology and landscape. Red Marl = Mercia Mudstone.

The ridge of Upper Greensand on the west side of the Sid Valley is called East Hill and the form of the base of the Upper Greensand is quite different on the west and east sides of the ridge. The west side is very straight for a distance of at least 5km, possibly because of an underlying fault. In contrast, the east side of the ridge is deeply intended by little tributaries of the River Sid that have cut back into the Upper Greensand. The Upper Greensand of East Hill is slightly tilted to the east, causing underground water to flow more readily to the east and emerge at the heads of the little tributaries mentioned above. East of East Hill, faults have isolated several small outlying hills of Upper Greensand from the main ridge – such as Path Hill [SY 137 935], Evergreen Hill [SY 136 929] and Sidbury Castle [SY 128 913]. The defensive advantage of these hill sites led to the larger of them (Sidbury Castle) being used by prehistoric man as the site of a hillfort.

Below the surrounding greensand scarp, the sheltered valleys are underlain by the red Triassic mudstone and most of this area is farmland. Many of the farms (whose names contain the word 'combe') are sited at the spring line at the base of the greensand to provide a guaranteed water source. The upper slopes of the valleys have many small pits which were dug in the past for agricultural marl. The distribution of these marlpits in an area on the east side of the Sid Valley is shown on the map on page 56.

The cliffs that contribute so much to the scenic attraction of the area are made up mainly of the eastward-dipping red Triassic rocks, overlain by yellowish brown Upper Greensand, white Chalk, and orange-brown Clay-with-flints. The contrast in colours of the formations making up the cliffs, as well as the differing reaction to erosion of the different types of rock, all contribute to the attraction of the coastal scenery.

At the coast, the landscape is sharply truncated. The generally straight trend of the coastline is sharply broken at Beer Head where the coastline turns abruptly northwards at nearly 90 degrees. This change of direction is probably related to the presence of north-south fractures associated with the major fault at Seaton Hole.

Between Otterton Ledge and Ladram Bay there are only small patches of shingle and fallen rocks at the base of the Otter Sandstone cliffs. Small caves have developed at the base of the cliffs. Below High Peak (west of Sidmouth) there are examples of rock platforms in the Otter Sandstone. East of there, most of the coast as far as Beer Head is fringed by a shingle beach. Patchy shingle is present between Seaton and Lyme Regis.

The cliffs in general are in a state of retreat caused by waves attacking their bases. The partial destruction of the coastal hillforts at High Peak and Berry Camp show how much cliff has disappeared in the past four thousand years. The rate of cliff retreat seems to be related to the presence or absence of shingle protecting the cliff base from waves. The location and thickness of the shingle can change dramatically according to local tides and currents. The movement of the shingle is governed by a process called longshore drift, which is the along-shore transport of shingle and sand by waves and currents. In our area this is mainly from west to east, so that in the west of our area (between Budleigh and Ladram Bay) the main pebbles in the shingle are quartzites derived from the Budleigh Salterton Pebble Beds. Eastwards from Ladram Bay [SY 096 852], chert and flint derived from the Upper Greensand and Chalk are the most common pebbles found in the shingle beaches. The shingle bars began to grow across the mouths of the rivers Otter, Sid and Axe at about the beginning of the 15th Century, causing silting and hindering access to the rivers by boats.

Sidbury Castle Iron Age hillfort from the south-west, with Burscombe Farm in the middle distance. The hillfort occupies the top of an outlying patch of Upper Greensand capped by Clay-with-flints.

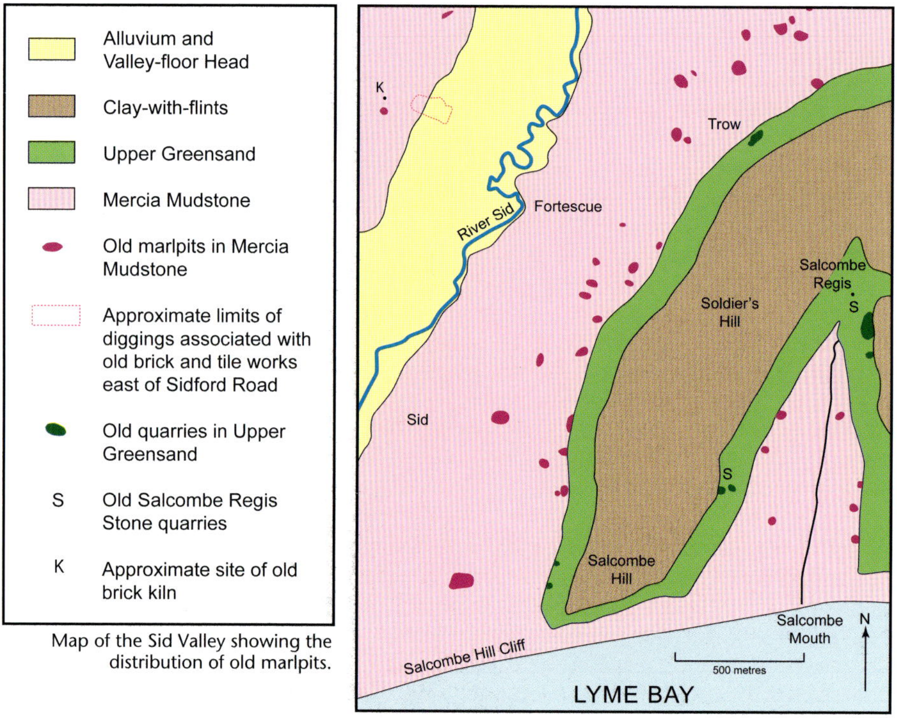

Map of the Sid Valley showing the distribution of old marlpits.

Because of the rapid retreat of the cliffs, smaller streams such as those along the Salcombe Valley and the Weston Valley have not been able to cut down their stream beds as fast as the rate of cliff retreat. Consequently, their valleys 'hang' above the shore. The larger River Sid has a wide valley out of proportion to its present size and erosive power. It is possible that the Sid carried a much greater volume of water at times during the Quaternary.

The middle and upper parts of the Upper Greensand (Whitecliff Chert plus Bindon Sandstone) is generally the toughest part of the succession and forms almost vertical cliffs. Between Beer Head and Beer, and east of Seaton, the Chalk and Upper Greensand together form steep high cliffs. A feature of the response to erosion of the Chalk cliffs around Beer is the contrast between the relatively unaltered chalk and a layer of partly dissolved chalk in the upper cliff; this upper layer is weaker and more prone to landslipping (see pages 44-45).

Between Seaton and Lyme Regis, the coastal scenery is dominated by major landslips in the area forming the Undercliffs Site of Special Scientific Interest. It is one of the largest active coastal landslip systems in Western Europe. The Great Landslip of 1839 was famous in its time and the first to be scientifically studied. It formed a huge block of land, 'Goat Island', behind which was left a deep cleft, 'The Chasm'.

A JOURNEY THROUGH THE ROCKS AND LANDSCAPES:
EXMOUTH TO PINHAY BAY

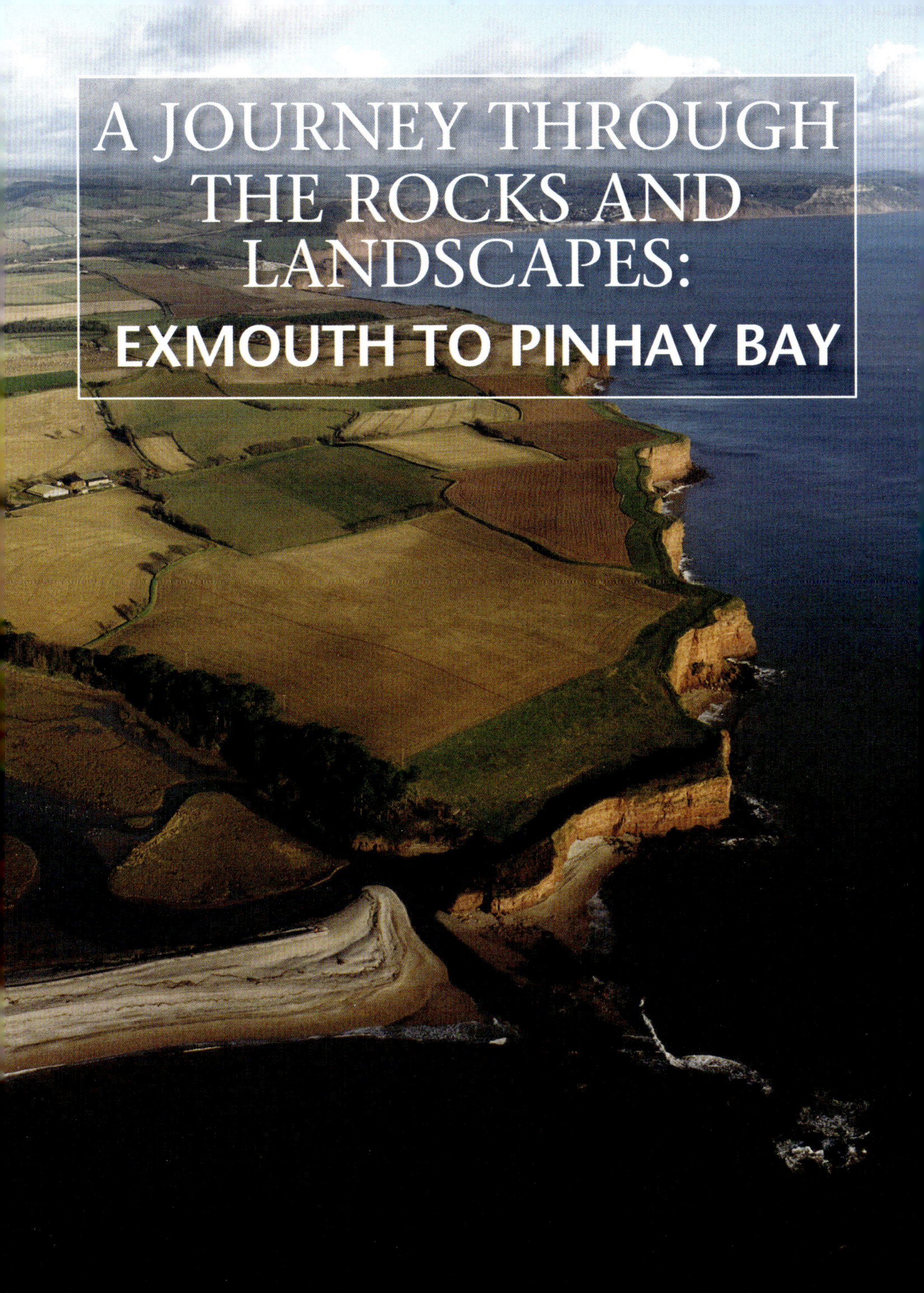

Aerial view of the mouth of the River Otter and Otter Sandstone cliffs to the east

In the last chapter we looked in a general way at the rocks, geological history and landscapes of the Red Coast. Now it is time to look in more detail at what can be seen along the coast, in the form of a journey of exploration from Exmouth in the west to Pinhay Bay in the east. The coast is divided into sections, with a summary of the geology and landscape and places of interest, some of which are described in greater detail in other sections of the book. The descriptions are given for exploratory walks both along the beach and along the clifftop. Details of access are given at the beginning of each section.

Exmouth to Budleigh Salterton

Like other seaside towns, such as Sidmouth and Budleigh Salterton, Exmouth gained a reputation as a resort during the Napoleonic Wars. It has since developed into a family holiday centre, aided by its sandy beaches which contrast with the shingle beaches that begin at Budleigh Salterton and extend eastwards into Dorset.

Access and walks

This section of coast is easily accessible from either Exmouth or Budleigh Salterton and there are convenient car parks at either end of this section. About midway there is parking at the Devon Cliffs Holiday Park at Sandy Bay, reached via Littleham. At low tide it is possible to walk along the sandy beach from Orcombe Point to Sandy Bay, returning either along the beach or taking the clifftop path back to Exmouth. For a walk from Exmouth to Budleigh Salterton a car can be left in Exmouth and a bus taken from Budleigh Salterton back to Exmouth.

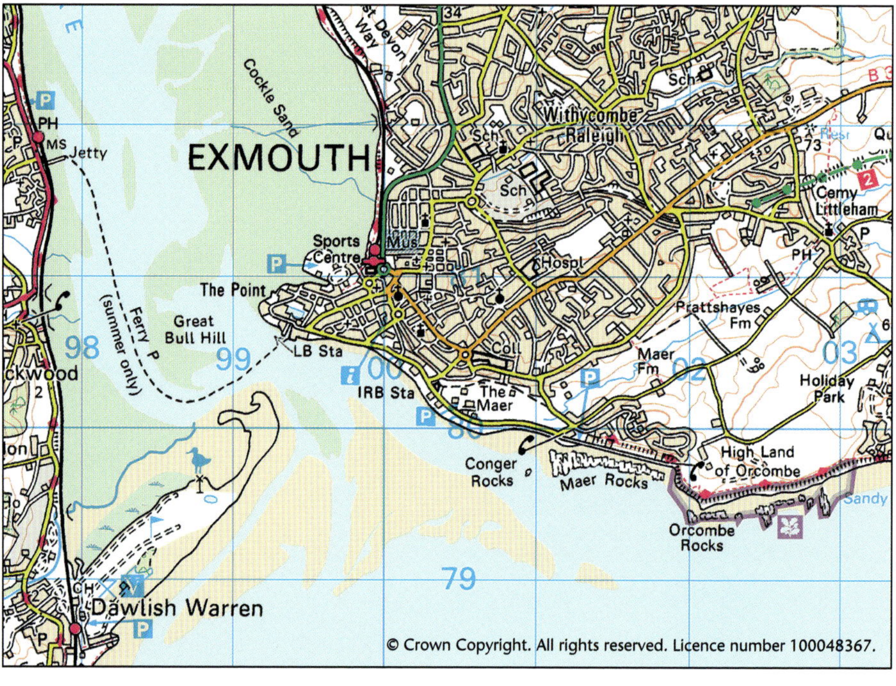

Beach exploration

Exe Breccia behind Exmouth Pavilion.

Madeira Walk extends south-eastwards from near the Exmouth Tourist Information Office [SY 000 807], through shrubberies and then continues along the back of The Maer (an area of old sand dunes). The path follows the base of an old cliff line marking the former position of the coast.

Starting in the west near the Tourist Information Office, the first rocks occuring in small cliffs about 7m high are breccias belonging to the Exe Breccia (see table on page 13). These breccias, described on page 32, are of Permian age, and the oldest rocks seen on our journey.

There are about 5m more breccia behind Exmouth Pavilion [SY 002 805]. However, on the south side of Carlton Hill Road, red clay and mudstone represent the Aylesbeare Mudstone that overlies the Exe Breccia. These mudstones, described on page 32, form the cliffs from here all the way to Budleigh Salterton.

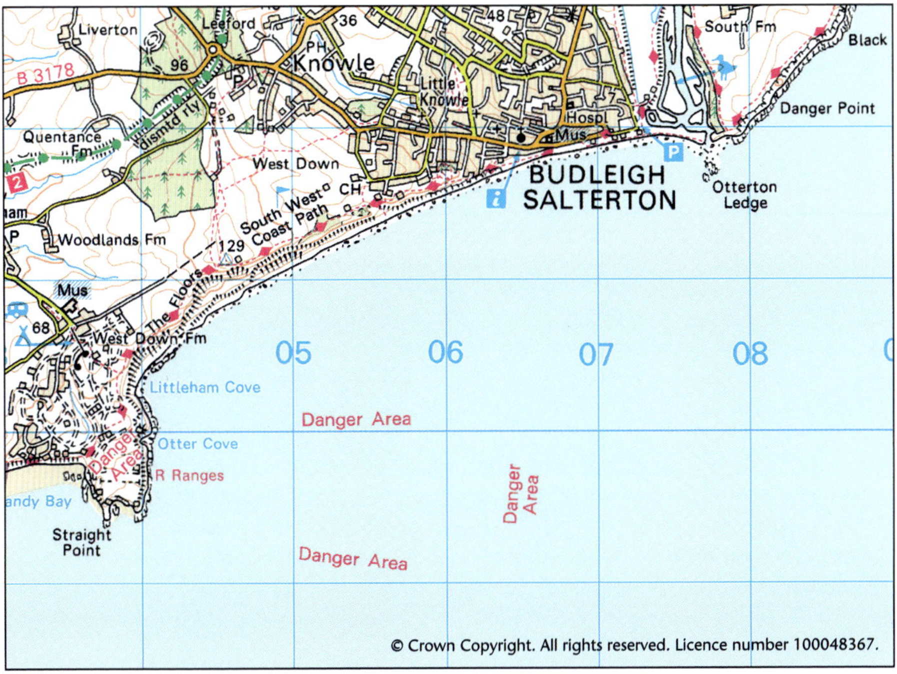

Progressing eastwards along Madeira Walk, there are about 2m of reddish brown sandstones on the north side of The Maer [SY 008 803], typical of those in the lower part of the Aylesbeare Mudstone (Exmouth Mudstone and Sandstone subdivision).

Along the sea front, extending for about 1km west of Orcombe Rocks, extensive reefs of mudstone called Maer Rocks are exposed at low tide. The cliffs north of Maer Rocks are largely vegetated but probably consist mainly of red mudstones. The World Heritage Site proper begins at Orcombe Rocks and is marked by a monument (the 'Geoneedle') on the clifftop above Orcombe Point.

Rodney Point and Orcombe Point just east form a prominent headland at the eastern end of the seafront road. East of Orcombe Point, the upper half of a small headland is formed mainly of red mudstones with a few thin sandstones, while the lower half is mostly red sandstone.

Immediately round the corner eastwards from Orcombe Point, near the steps down to the beach from the headland, the cliff shows a good example of a fault where movement has taken place on either side of a crack in the geological layers.

East of Orcombe Point, a headland formed of Aylesbeare Mudstone: mainly mudstones in the top half, sandstones below.

By comparing the position of a sandstone bed on each side of the fault, it can be seen that beds on the eastern side have moved down by about 2m relative to those on the western side. This displacement is called the throw of the fault – see the illustration on page 38 for an explanation.

A good example of a fault affecting the Aylesbeare Mudstone below the High Land of Orcombe, Sandy Bay.

About 200m east of Orcombe Point another fault brings massive red sandstone against red mudstone. The course of this fault can be traced as a rib on the foreshore running out to sea.

The eastward dip takes the sandstones of Orcombe Point to beach level just east of some steps. From here eastwards to Straight Point (about 1.5km) the cliffs above Sandy Bay are made up of red mudstones with a few thin beds of sandstone, representing the lower 200m or so of the Aylesbeare Mudstone (called the Exmouth Mudstone and Sandstone - see right). The beach is mainly sand with some reefs at low tide.

A view of a section of the cliffs in Sandy Bay between Orcombe Point and Straight Point, showing Aylesbeare Mudstone (mainly eastward-dipping red mudstone with some thin sandstones).

In the cliff below the High Land of Orcombe, there is a prominent sandstone near the top of the cliff, up to 2m thick but thinning out laterally. About 4m below is another sandstone which is thinner (about 0.5m) but more persistent than that above. Many large blocks from both sandstones have fallen onto the beach below. The cliff shows a good example of a fault which displaces the sandstones about 8 to 10m down to the east.

The eastward dip brings the prominent sandstone to beach level about halfway between Orcombe and Straight Points, to form a north-south reef, seen at low tides. East of this reef as far as Straight Point, the cliffs are largely mudstone with a few thin pale bands and some small faults.

At the eastern end of Sandy Bay, the east-west trending mudstone cliffs are abruptly terminated by the prominent north-south headland of Straight Point which owes its existence to the fact that it is partly made up of sandstone. These Straight Point sandstones are about 10m thick and can be traced inland about as far north as the village of Aylesbeare. There is a very large caravan site at the eastern end of Sandy Bay, and a firing range on the headland itself. It is not possible to walk around Straight Point at any time.

East of Straight Point, the coastline makes a re-entrant to form Littleham Cove. Radioactive nodules generally less than 7cm across were first found here by Mr Carter in the 1930s, and have since been found at other places inland. They are of particular interest because they are rich in vanadium, copper, uranium, nickel, cobalt, arsenic and silver.

The cliffs between Littleham Cove and Budleigh belong to the upper part of the Aylesbeare Mudstone (called the Littleham Mudstone), and consist of about 200m of red mudstones with some persistent thin beds of green silty sandstone.

At West Down Beacon, the Budleigh Salterton Pebble Beds that overlie the Aylesbeare Mudstone make their appearance at the top of the cliff. Their presence has an influence on the stability of the mudstone cliffs, for water from the porous pebble beds soaks into the underlying mudstones making them unstable. This gives rise to the landslips at The Floors. The general eastwards dip gradually brings the Pebble Beds lower down until, together with the overlying Otter Sandstone, they reach beach level at Budleigh Salterton.

Clifftop exploration

The clifftop from Orcombe Point to Budleigh Salterton makes a pleasant walk. A circular walk from Orcombe Point to the eastern end of Sandy Bay can be made with one section along the beach and the other along the clifftop, but careful attention needs to be paid to the state of the tide; it is best to do the beach part first to be certain of rounding Orcombe Point.

From the eastern end of the seafront road, near Rodney Point, the clifftop is reached via a zig-zag path. Following the path southwards, the clifftop above Orcombe Point, has a notable landmark, the 'Geoneedle', erected to mark the western end of the World Heritage Site.

Straight Point: sandstones with some mudstones, in the middle part of the Aylesbeare Mudstone.

The clifftop path rises steadily eastwards towards the High Land of Orcombe, from where there are good views westwards to the mouth of the Exe Estuary, Dawlish Warren, Teignmouth and beyond. In the top of the cliff are old river terrace gravels representing former deposits of the River Exe. They consist mainly of pebbles derived from the Budleigh Salterton Pebble Beds.

The 'Geoneedle' at Orcombe Point marks the western end of the World Heritage Site. The main part of the Geoneedle is made of Portland Stone, with inset stone panels indicating different types of rock found within the Site.

East of the High Land of Orcombe, the path descends to the large caravan site at Sandy Bay, and views of Straight Point open up. Cross the caravan site and regain the clifftop above Littleham Cove. From here follow the path eastwards and climb the escarpment of the Budleigh Salterton Pebble Beds at West Down Beacon. Here there are typical rounded Budleigh Pebbles in the soil. There is a seat at the Beacon with excellent views. From the Beacon there is a steady descent eastwards into Budleigh Salterton, partly beside a golf course. The cliffs are prone to instability, and a cliff fall in October 2001 caused the coastal path to be closed at [SY 0604 8166]. In places, gravels representing old terraces of the River Otter can be glimpsed in the clifftop.

Budleigh Salterton

The town of Budleigh Salterton lies on the coast about halfway between Exmouth and Sidmouth. Early sources give the name as Saltre in 1210 and Salterne in 1405 – these names derive from the salt pans (salterns) formerly worked near the mouth of the River Otter, on the east side of the town (see page 115). The main occupations up to the end of the 18th Century were salt-working, fishing and smuggling (note the name of Brandy Head on the cliffs east of the mouth of the Otter!). The town became a fashionable resort during the Napoleonic Wars when Europe was closed to travellers, and has retained its charm to the present day. It has a number of attractive Georgian and Regency buildings. There is a small museum, with an Earth Sciences exhibit, in a thatched building (Fairlynch House) in Fore Street.

Before the shingle ridge grew across the mouth of the estuary to almost close it off, it was deep enough to allow the passage of small ships. At the end of the 17th Century there are reports that the mouth of the river was wide enough to allow ships of sixty tons to work in *'as far as the upper salmon pool where there was twelve to fourteen feet of water...'*

Access and walks

The cliffs west of Budleigh can be reached from the town centre car park, or there is limited parking closer to the sea front in side streets such as Cliff Terrace. From there, the beach is reached via Steamer Steps.

The sea front east of the town is accessible from the large Limekiln car park [SY 073 821] (named from the former kilns sited there - see page 113) on the eastern side of the town, immediately next to the estuary. There is a path along the side of the shingle ridge to the mouth of the river (accessible for wheelchairs). There is no bridge across the mouth of the River Otter to the coast path east of the river. Instead, detour inland for about 1km to cross the river at White Bridge near South Farm, then turn southwards along the east side of the estuary for about another 1km before regaining the coast.

A popular level walk (about 3km) from the Limekiln car park is northwards along the River Otter to the attractive village of Otterton, which has a working flour mill. The Domesday survey recorded a mill with four stones, the same as it still has – a remarkable record of continuity. Along the way you can see low river-cliffs of red Otter Sandstone on the east side of the river. On the west side of the path, the flat land was formerly saltmarsh, reclaimed for agriculture by the construction of banks to keep out the sea. These banks were built by French prisoners during the Napoleonic Wars. Other walks in the Otter Valley are included in a booklet published by the Otter Valley Association (*Twelve walks in the Otter Valley*).

West Cliff, Budleigh Salterton

West of Budleigh Salterton, at West Cliff [SY 062 816], there is a distinctive layer made up of well-rounded pebbles. This layer, about 30m thick, is the celebrated Triassic Budleigh Salterton Pebble Beds. The photograph below shows a general view of the Pebble Beds and the overlying Triassic Otter Sandstone (with distinctive honeycomb weathering).

Because of their unique appearance and the fossils that can be found in the pebbles, the Pebble Beds are probably the best known and most widely studied of the New Red Sandstone formations of Devon. They represent the deposits of ancient (about 245 million years old) rivers that flowed from the south. More details about the Pebble Beds can be found on pages 32-33.

A general view of the Budleigh Salterton Pebble Beds and the overlying Otter Sandstone at West Cliff, Budleigh Salterton. Note the conspicuous honeycomb weathering of the Otter Sandstone.

As the Pebble Beds are traced eastwards in the cliff from West Down Beacon, an easterly tilt of about 5 degrees brings them down to beach level at West Cliff. They overlie red mudstone of the Aylesbeare Mudstone very sharply, but in the cliffs the boundary is often covered by slipped material.

A layer of ventifacts (see page 33) occurs at the top of the Pebble Beds in the cliff face west of Budleigh Salterton, and they have also been seen on the surface at Woodbury Common. The ventifacts are associated with a thin layer, about 0.3m thick, of red clay representing a fossil soil.

The shingle beach at Budleigh Salterton is mainly made up of pebbles which have fallen from the Pebble Beds. The beach pebbles have been shaped by the grinding action of the waves giving them a discoidal shape compared with the more rounded shape of those in the cliff (see photo, page 66). Most of the pebbles consist of a very hard type of sandstone called quartzite. They are mostly grey, sometimes with red mottling. Other types include porphyry, a type of fine-grained volcanic rock with crystals of the mineral feldspar.

Some of the quartzite pebbles on the beach have rusty-looking cavities that sometimes indicate the presence of fossils when the pebble is broken open. They are not common, but diligent and patient search by Victorian collectors yielded up a wide variety of fossils, mainly brachiopods, bivalves and trilobites. The drawings on this page show some of these fossils. The fossils are not contemporary with the Pebble Beds, but are

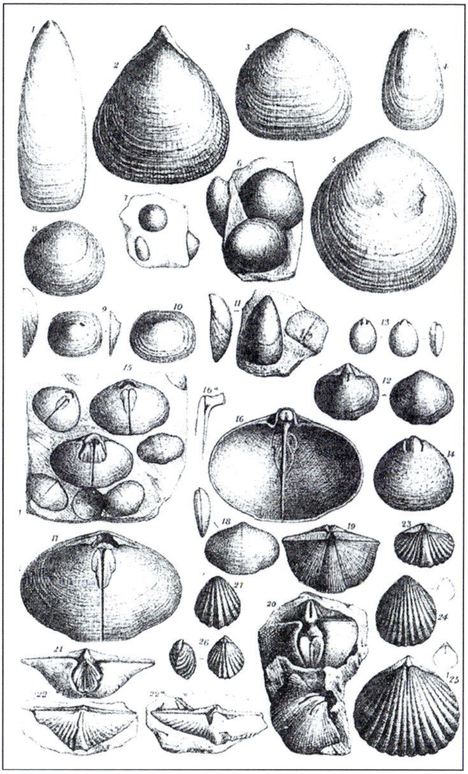

Brachiopod and bivalve fossils collected by Victorian naturalists from the quartzite pebbles of the Budleigh Salterton Pebble Beds.

Detail of the boundary between the Budleigh Salterton Pebble Beds and the overlying Otter Sandstone at West Cliff, Budleigh Salterton. The boundary is marked by a layer of polished and faceted pebbles (ventifacts) which overlie a fossil soil layer.

derived from much older rocks that were later worn down and washed into the Pebble Beds rivers.

Detail of the shingle at West Cliff, Budleigh Salterton. Rounded pebbles from the Budleigh Salterton Pebble Beds have been reworked into the beach shingle where they have acquired a discoidal shape by the grinding action of the waves.

Still at West Cliff, examine the layer that overlies the fossil soil and ventifact layer. There is a marked change to the sandstones of the Otter Sandstone, (see the photos on pages 64 and 65). The lowermost 10-15cm of the Otter Sandstone is a conspicuous yellow colour. Most of the Otter Sandstone was formed in a large sandy river that, like that depositing the Pebble Beds, also flowed from south to north. However, the lowest 14m immediately overlying the Pebble Beds at Budleigh Salterton were laid down as wind-blown sand.

Calcareous concretions in the Otter Sandstone at Budleigh Salterton sea front, that probably formed around plant roots.

Budleigh sea front

Continue the eastwards journey, along the sea front at Budleigh Salterton towards the mouth of the River Otter. There is a gap of about 0.9km before the next Otter Sandstone is visible, in low cliffs [SY 070 819]. They locally show honeycomb weathering. In contrast to the sandstones immediately above the Pebble Beds, these sandstones were laid down by rivers. Notable features here are well-developed subvertical calcareous concretions. They represent deposition of calcite around the roots of plants that grew during Otter Sandstone times, so that the concretions mimic the subvertical cylindrical shapes of the roots.

In the road above the seafront, at Salting Hill, a wall [SY 0714 8198] shows interesting use of the local building materials. It is built of evenly-sized cobbles of quartzite from the Budleigh Salterton Pebble Beds, with an upper course of irregular blocks of conglomerate from the Otter Sandstone.

The shingle ridge, composed mainly of quartzite pebbles, extends along the sea front all the way to the mouth of the River Otter, where the river squeezes out to sea through a narrow gap between the ridge and the Otter Sandstone cliffs.

Immediately behind the shingle ridge is a wide flat area of saltmarsh deposits, well seen from the Limekiln car park [SY 073 821] where there is an East Devon Heritage Coast information board. The saltmarsh is rich in bird-life, which can be observed from easily accessible hides on both sides of the estuary. The splendid panoramic aerial view (on pages 10-11) shows the shingle ridge across the mouth of the river, the saltmarsh behind, and the red Otter Sandstone cliffs. In the far distance on the right, High Peak can be made out.

Otterton Ledge to Ladram Bay (about 4km)

The stretch of coastline between Otterton Ledge [SY 077 819] at the mouth of the River Otter, and Ladram Bay [SY 096 852] is made up of near-vertical cliffs of Otter Sandstone up to about 60m high. The coastline is fairly straight, with a few small headlands such as Danger Point, Black Head, Brandy Head and Smallstones Point, until Ladram Bay is reached. At Ladram Bay, the Otter Sandstone has been eroded into a series of impressive sea-stacks.

A wall at Salting Hill, Budleigh Salterton, showing the use of well rounded pebbles from the Budleigh Salterton Pebble Beds in the lower half and blocks of conglomerate from the Otter Sandstone in the upper half.

View of the mouth of the River Otter. The river reaches the sea through a narrow gap between the end of the shingle ridge and cliffs of Otter Sandstone. The shingle ridge is composed of pebbles from the Budleigh Salterton Pebble Beds and is banked up against the reefs of Otterton Ledge, which can be seen extending out to sea.

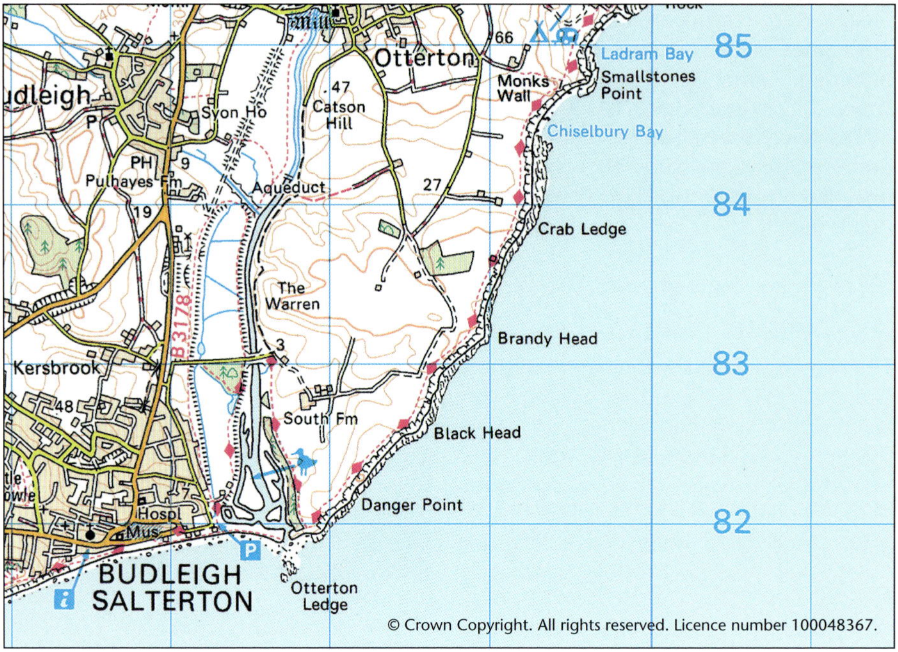

Access and walks

The South West Coast Path follows the top of the cliff from the mouth of the River Otter to Ladram Bay. There are car parks at Budleigh Salterton [SY 073 821] and at Ladram Bay [SY 096 853]. The walking is generally easy, with none of the major ups and downs that, as we will see later, are so characteristic of the coast east of Ladram Bay. The Otter Valley Association booklet *Twelve walks in the Otter Valley* gives details of several circular walks, including this part of the coast.

Beach exploration

The base of the cliff is generally not easily accessible, except at Ladram Bay [SY 096 852]. At Danger Point [SY 081 821] are examples of the few faults that affect this part of the section. The best way to see the rest of the section is from a boat, from which can be seen the general character of the sandstones, including their general eastward inclination (at around 5 degrees). Many small caves can be seen at the base of the cliffs. The cliffs are subject to rockfalls of various sizes, and there are small patches of shingle in some of the coves at the cliff base.

Faults affecting the Otter Sandstone near Danger Point indicated by broken white lines.

Clifftop exploration

Archaeological excavations at the clifftop site just east of the mouth of the River Otter revealed a possible Roman villa site from the late second and third centuries.

A walk along the clifftop path between Otterton Ledge and Ladram Bay reveals glimpses of the cliffs of Otter Sandstone. A good viewpoint is [SY 079 820], looking eastwards towards Danger Point (see photo on page 27).

In the clifftop patches of river terrace gravels laid down by the River Otter overlie the sandstones. They are made up of pieces of chert derived from the Upper Greensand, and rounded pebbles derived from the Budleigh Salterton Pebble Beds. In the clifftop about 250m north of Brandy Head [SY 0895 8333] are about 2m of inaccessible river terrace gravel. In the clifftop above Twopenny Loaf Rock [SY 0905 8363] there are over 2m of inaccessible river terrace gravel. At the south end of Chiselbury Bay, coastal erosion has cut into the river terrace gravels, and about 3.5m of flinty gravel and sand can be seen in the clifftop [SY 0923 8398].

Approaching Ladram Bay, there is a splendid view to the north-east from a viewpoint [SY 094 848], near the site of the Vat Well. The sea-stacks of Ladram Bay can be seen in the foreground, with High Peak above; Sidmouth and Salcombe Hill can be seen in the mid-distance, and beyond Sidmouth the flat-topped coastline stretching to the white Chalk cliffs of Beer Head.

The Vat Well is a spring sited at the junction of the Otter Sandstone with the overlying Mercia Mudstone.

A view from south of Ladram Bay looking north-east.

Ladram Bay

Ladram Bay [SY 096 852] is a popular place with summer visitors. It is a good place to look more closely at the Otter Sandstone, and also to see the sea-stacks which are such a characteristic feature of this charming place.

The main NE-SW-trending cliff face, about 200m long, shows the Otter Sandstone to consist mainly of reddish brown sandstones, with a few impersistent layers of reddish brown mudstone up to 0.25m thick. The sandstones are divided up by several laterally persistent levels which represent erosion surfaces formed by ancient rivers. The rocks therefore consist of a stack of river channels preserved in the rocks. These channels have in places cut down into mudstone layers, and at the channel bases are 'lag' deposits of conglomerate. Cross-bedding in the sandstones represents the migration of sand within the river channel. Its orientation indicates the direction in which water flowed in the channels, which here and at most other Otter Sandstone localities, is from south to north. The rarely preserved mudstones represent the muddy infill of abandoned river channels.

The sea-stacks at Ladram Bay are textbook examples. Some of them – for example Hern Point Rock [SY 0992 8534] are capped by small pinnacles of Mercia Mudstone. What is the reason for the proliferation at stacks at Ladram Bay, but not along the coastline farther south-west? It is probably partly because faults and joints are commoner at Ladram Bay, causing lines of weakness which have been exploited by the sea. Also, because hard beds provide a rock platform in the intertidal area on which the stacks sit, they are better able to

Formation of sea-stacks

Stage 1
An arch forms in the Otter Sandstone as the sea works into weaknesses in the rock.

Stage 2
The arch gradually enlarges as more and more sandstone is eaten away.

Stage 3
The arch eventually collapses into the sea, leaving the seaward pillar as an isolated stack.

resist wave erosion at their bases. The classical development of a sea-stack begins with a headland, and progresses by continued wave attack, via the formation of a natural arch which then collapses, to an isolated stack. For example, in 1925 the arch at Ladram Bay collapsed, which then isolated the sea-stack.

Good examples of shore platforms formed by more resistant layers of conglomerate within the Otter Sandstone are found in the Ladram Bay area, for example at the southern end of the bay around Smallstones Point. The cliffs are also notable for examples of honeycomb weathering caused by wind and salt erosion of the sandstones, which etch out the sedimentary structures in the rocks.

In contrast to the beach farther west at Budleigh Salterton, the shingle beach at Ladram Bay is mainly made up of flints.

Hern Point Rock. Mercia Mudstone-capped pinnacles on sea-stacks of Otter Sandstone at Ladram Bay.

Ladram Bay to Sidmouth (about 3km)

Along this section of the coast there are three new geological units: the Mercia Mudstone that overlies the Otter Sandstone; the Upper Greensand that forms the hills of High Peak and Peak Hill; and the Clay-with-flints that caps the Upper Greensand at Peak Hill.

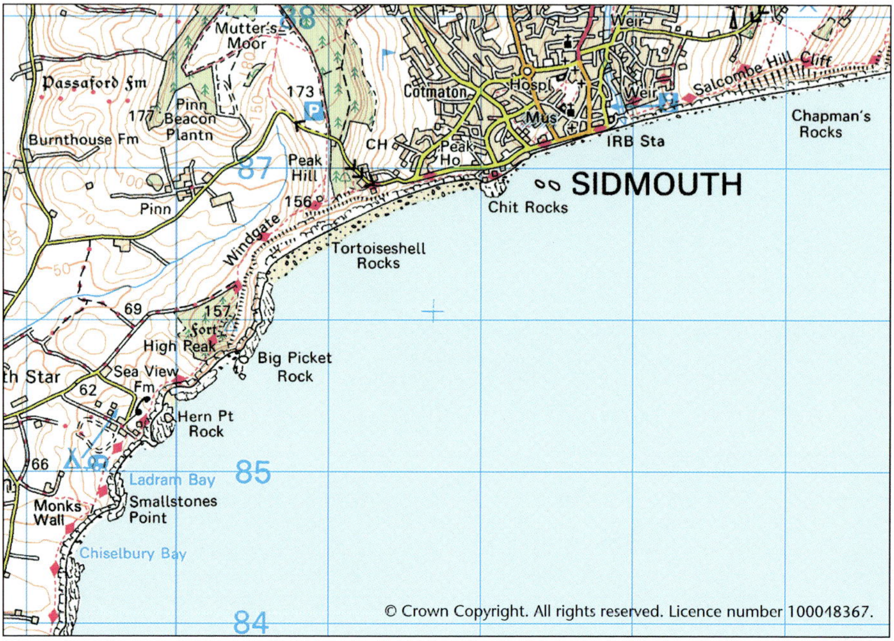

View westwards from Jacob's Ladder towards High Peak and Peak Hill. In the distance, the cliffs below High Peak show Otter Sandstone at the base, overlain by a less steep cliff profile of Mercia Mudstone. The top of High Peak is capped by wooded Upper Greensand (UGS). East of High Peak the eastward-dipping Otter Sandstone disappears beneath beach shingle below Peak Hill. The cliffs below Peak Hill are formed mainly of Mercia Mudstone. The vegetated top of the cliff is Upper Greensand capped by Clay-with-flints. In the near-distance, the cliffs are of Mercia Mudstone, but the dip has been reversed and is to the west.

Access and walks

There is easy access to this section of the coast from Ladram Bay at the western end, where there is a car park, caravan site, shops and a pub. Peak Hill is easily accessible by a short level walk southwards for about 500m from the car park [SY 109 873] beside Peak Hill Road. A leaflet giving details of walks in the Peak Hill area is published by the Sid Vale Association. At the Sidmouth end of this section of the coast there is a large car park [SY 120 871] at Connaught Gardens on the western side of Sidmouth, convenient for access to the beach west of Jacob's Ladder.

Beach exploration

Please note that the stretch of coast from High Peak to Beer Head is especially subject to frequent cliff falls and great care should be taken – do not approach the cliffs too closely, or attempt to climb them. The state of the tide also needs to be considered when planning an excursion. Fortunately the main features of interest can be appreciated from a safe distance.

It is not possible - without getting wet! - to get from Ladram Bay eastwards into the next bay round Ladram Arch, so the best option is to walk westwards for about 2km from Jacob's Ladder to the north side of Ladram Arch. It is very important to have low tide for this excursion. The beach is shingle and slow going, but at low tide there are areas of easier walking along sand exposed on the seaward side of the shingle. A good start before beginning is to go to the top of Jacob's Ladder [SY 1200 8694] and look westwards towards the cliffs beneath Peak Hill and High Peak. The photograph below shows this view, with annotations explaining the different geological units.

The coves around Hern Rock, and Sandy Cove immediately to the north are beautiful places at low tide, with a lovely cut-off feeling. The sea-stacks sit on a rock platform

of seaweed-covered Otter Sandstone conglomerate with patches of sand. The vertical amphitheatre-like cliffs above these two coves show Otter Sandstone, overlain by Mercia Mudstone at the clifftop. A notable feature in Sandy Cove is a large fault which brings Otter Sandstone against Mercia Mudstone at the clifftop, indicating that the south side of the fault has moved downwards relative to the north side. Other faults are also visible in the cliff faces. Follow the rock platform round to the sea-stack of Big Picket Rock [SY 1044 8574], composed of red sandstone. This is a good place to see the curious calcareous concretions, many vertical, which may have formed by deposition of calcite around plant roots (see the photograph on page 66).

Shingle west of Jacob's Ladder, Sidmouth, showing mainly flint pebbles.

At Lade Foot Cove [SY 106 862] there is a fault in the cliff. Also in this area are caves which have been excavated in the base of the cliff by wave action. North of the cove, the rocky shore platform ends, and the foreshore is made up of shingle all the way to Sidmouth, with some sand at low tide. In contrast to the shingle at Budleigh Salterton, this shingle is made up almost entirely of flint pebbles (above).

The boundary between the pale grey sandstone of the Otter Sandstone at the base of the cliff and the overlying red Mercia Mudstone below Peak Hill. The lower part of the Mercia Mudstone contains thin layers of green mudstone and lines of small cavities where nodules of gypsum have weathered out of the mudstone.

Farther along the Coast towards Sidmouth, the Otter Sandstone is overlain by the Mercia Mudstone, with the boundary between the two gradually dipping eastwards at about 2 degrees until it reaches beach level below Peak Hill. The boundary is well defined by a sandstone bed at the top of the Otter Sandstone (see opposite). This bed is rather variable in thickness, pinching and swelling in places to almost nothing, but is still persistent.

Eastwards from the place where the Otter Sandstone disappears below beach level beneath Peak Hill, the cliffs are made up of red Mercia Mudstone, capped in the higher part of the cliff by the Upper Greensand and Clay-with-flints of Peak Hill. The Mercia Mudstone dips eastwards below Peak Hill at between 2 and 4 degrees, until an area about 600m west of Jacob's Ladder. At this point there is a small fault in the cliff, east of which the dips in the Mercia Mudstone are reversed, and dips are to the west at about one or two degrees. There is another small fault in the cliff about 230m west of Jacob's Ladder.

Clifftop exploration

From Ladram Bay there is a steepish walk up the slope towards High Peak. Not far up the slope above the pub (the 'Three Rocks') there are lovely views north-eastwards towards Sidmouth, with the sea-stacks in the foreground. The closest large stack is Hern Point Rock [SY 0992 8534] and the stack in the mid-distance is Big Picket Rock [SY 1044 8574]. Hern Point Rock, and the two smaller stacks just south of it, are capped by little pinnacles of Mercia Mudstone (see photo on page 71). Note also the profile of the cliff below High Peak. The lower half of the cliff is made up of Otter Sandstone, forming a very steep, almost vertical cliff. The upper half is made up of Mercia Mudstone, which is less resistant to weathering and gives rise to a less steep cliff profile. The vegetated top of the cliff is formed of a capping of Upper Greensand on top of the Mercia Mudstone.

The clifftop above Sandy Cove is made up of a thin layer of Mercia Mudstone overlying the Otter Sandstone. However a fault running east-west through the north side of Sandy Cove brings the Otter Sandstone in for a while until the main Mercia Mudstone outcrop begins north of Conger Pool. From here up to the base of High Peak, the soils are clayey, reflecting the underlying geology.

The coastal path now enters woodland and steers away from the cliff edge to skirt the base of the capping of Upper Greensand that forms High Peak. To visit High Peak, a path diverts eastwards from the main path to climb steeply to the hilltop. There is no problem recognising the base of the Upper Greensand where there is a sudden steepening of the slope. Once on the Upper Greensand yellowish brown sand ('Foxmould') is thrown out from burrows. The top of the hill, 157m high, and marked by a trig point, is an airy spot, with splendid views.

High Peak is notable also for being the site of a prehistoric hillfort, although much of it has been lost by erosion of the cliffs. Today it is difficult to appreciate the earthworks because of afforestation. The fort was defended on the landward side by a rampart scarp and ditch with a small east-facing outer rampart. The earliest evidence for the occupation of the site, based on finds of flint implements, pottery and radiocarbon dating, is in the Neolithic (fourth millennium BC), and later in the Roman period, when it was possibly in use as a coastal observation post. Interestingly, although the site was long thought to be an Iron Age fort, excavations in 1961 by the archaeologist Mrs Pollard suggested that the site was re-occupied and the present earthworks built during a period of Dark Age occupation. The discovery of imported Mediterranean wine amphorae suggests that the site may have been of fairly high status.

Descend from High Peak to regain the coastal path. Continuing towards Sidmouth, leave the woodland and descend into the area between High Peak and Peak Hill, known as Windgate. All this area is underlain by the Mercia Mudstone. Before leaving the woodland, there is a seat and viewpoint eastwards towards Sidmouth. Near Windgate, the path crosses a hollow [SY 1042 8634] – an example of the many disused marlpits which are scattered over the Mercia Mudstone.

On the other side of Windgate, there is a sudden upward steepening of the slope and a distinct change in land use from the cultivated fields on the Mercia Mudstone to rough grass and bracken developed on the Upper Greensand. Eventually the slope slackens and you are on the flat-topped part of Peak Hill. In the clifftop is a layer of brown Clay-with-flints overlying the Upper Greensand.

There are excellent views southwards to High Peak from here. Inland, the plateau of the Clay-with-flints stretches inland towards Mutter's Moor [SY 105 881] and Bulverton Hill [SY 107 890]. Mutter's Moor has yielded a scatter of flint tools and waste flakes (pages 104-105), suggesting that people lived here in Neolithic (New Stone Age) times.

Eastwards from Peak Hill, the coastal path enters an area of overgrown and uneven ground where the Clay-with-flints was once worked in old gravel pits, probably for repairing roads. Descend the steep Upper Greensand escarpment on the east side of Peak Hill. The coastal path meets the Otterton to Sidmouth road at the base of the

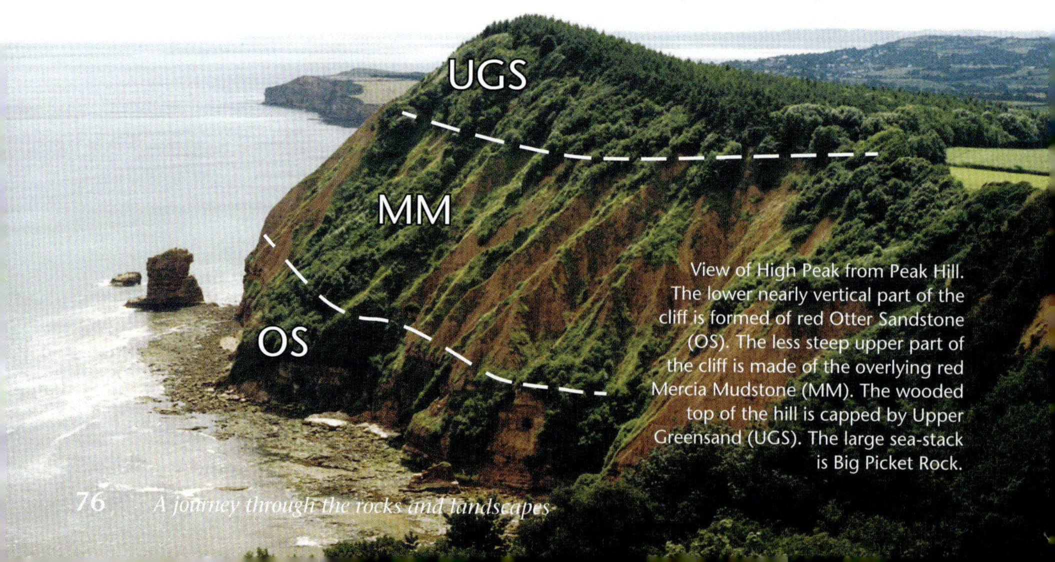

View of High Peak from Peak Hill. The lower nearly vertical part of the cliff is formed of red Otter Sandstone (OS). The less steep upper part of the cliff is made of the overlying red Mercia Mudstone (MM). The wooded top of the hill is capped by Upper Greensand (UGS). The large sea-stack is Big Picket Rock.

Upper Greensand [SY 1120 8688]. Further evidence of coastal erosion was seen in 1996 when the clifftop [SY 1133 8684] collapsed and the road was closed; it has now been rebuilt slightly farther inland. The descent towards Sidmouth is through a pleasant open field (Cliff Field) with many benches which are particularly welcome for those making the ascent from the east. There are good views ahead of Sidmouth and the cliffs beyond to Beer Head.

Sidmouth

Sidmouth, the Sedemuda (meaning the mouth of the river Sid) of the Domesday Book, is a small seaside town, unspoilt and retaining much of its early 19th Century character. It lies in a charming position spread out along a valley framed by hills: to the west, Peak Hill, and to the east, Salcombe Hill. For centuries, it was a small fishing and market town. In the early 17th Century Risdon described it as *'one of the especialest fisher towns of the shire'*. However, the decline of the fishing industry coincided with the discovery of the town as a resort with a pleasantly mild climate and beautiful scenery. The numbers of visitors increased from the late 18th Century, mainly owing to the closure of the Continent during the Napoleonic Wars. By 1796, the town was '...*of late tolerably frequented in the bathing season*' (*Copper Photo Magazine*, 1796).

There is an interesting museum, which includes geological exhibits, housed in Hope Cottage near the town centre.

Wealthier visitors to Sidmouth built 'cottages', many of which were later converted to hotels. The architecture is mainly Georgian and Regency, with elegant wrought-iron balconies, tent-shaped canopies and trellis supports. There are notable examples of the thatched cottage style called 'cottage orné, (e.g. Clifton Cottage at the western end of the Esplanade), and 'Strawberry Hill Gothic' with castellated battlements and Gothicised windows (e.g. Coburg Terrace and the Royal Glen Hotel). It was at the Royal Glen Hotel (formerly Woolbrook House) where Queen Victoria stayed as a child in 1819-20 with her parents the Duke and Duchess of Kent. Fortfield Terrace was begun in 1792 by a London speculator, although never completed as planned.

Jacob's Ladder [SY 1200 8694], a well-known Sidmouth landmark, is a set of white-painted wooden steps connecting the beach with the clifftop on which Connaught Gardens are set. Interestingly, there is a geological control to the siting of the Ladder. At this point a large north-south - trending fault extends out to the coast. On the west side is Mercia Mudstone and on the east side is Otter Sandstone; consequently, the west side has dropped relative to the east by an estimated 75m. A borehole just west of the base of the Ladder would encounter the Otter Sandstone again at an estimated depth of about 38m. The fault can be seen at low tide running out along the foreshore reefs. Those on the east, called Chit Rocks, are of sandstone. Chit Rock has an interesting history. It was formerly a sea stack, as prominent as those in the Ladram Bay area, but was entirely destroyed by the huge storm of 23rd November 1824. This was a particular loss to the town, for the Rock was venerated by the local fishermen, and there was an annual procession to the rock at low tide when one was chosen to be 'King of Chit' and was crowned and enthroned on the rock!

There were formerly limekilns at the top of the cliff where Connaught Gardens are now sited. The limestone was carried up to the kilns by donkeys using a track up from the beach.

A path at the base of the sandstone cliffs connects the base of Jacob's Ladder with the western end of Sidmouth Esplanade. The Otter Sandstone can be seen at close quarters along this path.

The town of Sidmouth is fronted by a shingle ridge. A rolled earth bank was built on it in 1805, but failed to protect the town from flooding. For example, during the great storm of 1824 referred to above, the sea broke through and caused widespread damage and the whole of the lower part of the town was flooded. The first permanent Esplanade was completed in 1837. Removal of the shingle in front of the sea wall by the action of the sea and by townspeople taking it for building purposes led to pressure on the foundations of the wall, and during the 1870s and 1880s little was done except to repair the surface of the Esplanade. In 1923 the lack of repair led to holes appearing in the Esplanade and the sea broke through and flooded the town. In 1924 a violent storm ripped up great lengths of the Esplanade. As a result a new Esplanade was built and opened in 1926. Bringing the story up to date, artificial islands were recently constructed in front of the town, designed to reduce the danger to the town in severe storms. They have radically altered the seaward outlook of the town. At the same time, large quantities of shingle were brought in from the Blackhill quarries in the Pebble Beds and used to form a new beach.

Beneath the shingle on the foreshore there are possible remnants of a 'submerged forest' (see page 51), formed when sea-level was lower than it is today. In 1873 the well known Sidmouth antiquary, Orlando Hutchinson recorded traces of submerged vegetation on the foreshore: stumps of trees were found 2.4m below high water mark. He recorded 'Blue clay' from under the shingle ridge, at about high water mark, and teeth of Woolly Mammoth *(M. primigenus)* were found in the clay. These specimens date from the last cold stage of the Quaternary, (see page 46) and can be seen in Sidmouth Museum.

Various schemes have been proposed - in 1811, 1825, 1836 and 1862 - to build a harbour for Sidmouth, but came to nothing because of shortage of funds. The position of the proposed harbours varied between the mouth of the river in the area of the Ham, and Chit Rocks. The third scheme, in 1836, was the most elaborate. The harbour was to be built at the western end of the front, in the Chit Rocks area, with two piers planned to enclose an area of 10 acres, with an opening for shipping on the south-east side. A large foundation stone was laid on Chit Rocks, but this was the full extent of the building works! The scheme by which stone was to be provided for construction proved to have a farcical ending. The stone was to be brought from Hook Ebb, about 2km east of the mouth of the River Sid. There is no suitable stone in the cliffs at Hook Ebb, the cliffs being made of Mercia Mudstone, so presumably fallen blocks from the Upper Greensand above were to be used. To get to Hook Ebb, a railway tunnel about 1/3 mile long was dug through the cliff parallel to it, then the line ran eastwards along the beach. Hutchinson, in his *History of Sidmouth*, writes that for many years afterwards the posts that supported the timbers could be seen poking out of the shingle at the base of the cliff. The mouth of the tunnel can be seen in an old photograph (right) but is no longer visible, and has probably been removed by erosion. Traces of the tunnel can still be found. A gated entrance about 300m east of the river mouth (below) is a possible side entrance to the old railway tunnel. The present position of the eastern exit of the tunnel could be seen at one time, although it was completely blocked by a cliff fall in 2001. A steam engine was acquired to run in the tunnel, and brought to Sidmouth by sea. However, a major miscalculation had been made, because the engine proved to be too large to fit in the tunnel! It was run along the Esplanade for joyrides for a while. The eventual fate of the locomotive is uncertain.

Former entrance (T) to the railway tunnel just east of the mouth of the River Sid, 1901.

At the eastern end of the shingle ridge, the River Sid is confined to a narrow space between the shingle ridge to the west and sandstone cliffs to the east, a situation similar to that at the mouth of the River Otter.

Sidmouth town is mainly underlain by Otter Sandstone. The sandstones were extracted from a small quarry [SY 1263 8869], now partly occupied by a clubhouse. About 2m of red sandstone can be seen dipping eastwards at 3 to 4 degrees. Another old sandstone quarry [SY 1263 8840], just east of Ascerton Close, is now occupied by a building.

A gated side entrance which possibly leads to the old railway tunnel in Salcombe Hill Cliff, east of the mouth of the River Sid.

The main mass of sandstone in Sidmouth town is abruptly cut off on the west side by a northward extension of the fault seen at the coast at Jacob's Ladder. The fault brings Mercia Mudstone on the west against Otter Sandstone on the east. It extends north from Jacob's Ladder to pass just west of the Council offices at Knowle, and then north to Bulverton and Higher Woolbrook. Another fault along the Wool Brook valley brings Mercia Mudstone on the north against Otter Sandstone on the south. Finally, to the east, the Otter Sandstone dips eastwards beneath Mercia Mudstone, the boundary between the two formations running along the valley of the River Sid.

Sidmouth to Branscombe Mouth (about 8km)

Between the mouth of the River Sid and Branscombe Mouth, the cliffs are similar in general geological make-up. Views eastwards from Sidmouth (see photograph on page 40) show a striking colour contrast reflecting the two main ages of the rocks in the cliffs; the red Triassic rocks form the lower half of the cliff; and Cretaceous rocks, comprising the pale yellow-brown Upper Greensand below and the white Chalk above, form the upper part of the cliff. The topmost layer is a thin skin of Clay-with-flints. This section of coast consists of a series of flat-topped hills cut into in three main places – at Salcombe Mouth, Weston Mouth and Branscombe Mouth – by steep-sided valleys or combes.

Access and walks

This stretch of coast can be walked in one fairly strenuous go, but selected parts can be visited from several intermediate access points. This has the advantage that the coast can be reached along mostly level paths without the ups and downs involved in walking the coast path. Some of the intermediate access points are as follows: The western part of the coast path (Salcombe Hill Cliff) is accessible from Sidmouth (there is a car park near the swimming pool on the east side of the town). Salcombe Hill is easily accessible by a level path from the National Trust car park [SY 139 882] near the Observatory. A leaflet published by the Sid Vale Association gives details of walks around Salcombe Hill. Salcombe Mouth can be reached from a car park [SY 148 888] near the church in Salcombe Regis. Higher and Lower Dunscombe cliffs can be reached by parking [SY 154 887] beside the Salcombe to Dunscombe road (limited parking) and walking south to the head of Lincombe where there is a choice of paths either west to Higher Dunscombe Cliff or east to Lower Dunscombe Cliff and Rempstone Rocks. Rempstone Rocks can also be reached more directly from Dunscombe through Dunscombe Coppice. Weston Mouth is accessible from a car park [SY 1661 8896] in Weston. Weston Cliff and Coxe's Cliff can be reached by a path that leaves the Weston to Branscombe road at [SY 173 889]. Berry Cliff is accessible by a path leading south from Branscombe Church. The eastern part of the section (West Cliff, just above Branscombe Mouth) is accessible by a steep ascent along the coast path from Branscombe Mouth, where there is a large car park.

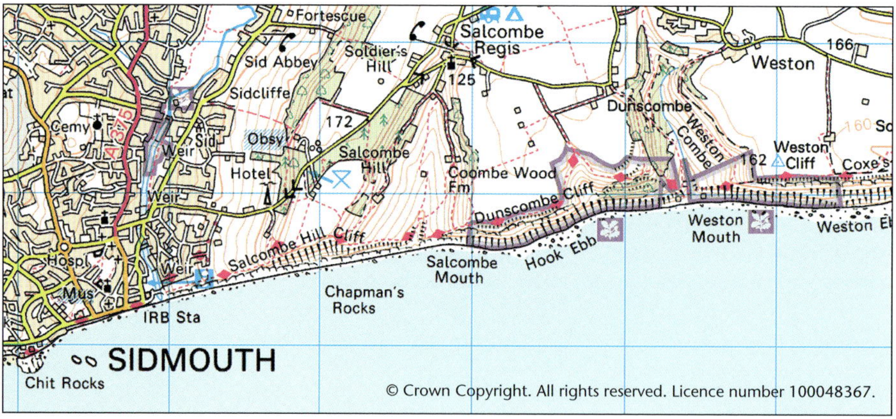

Beach exploration

If the tides are right, the lower cliff can be explored by walking from Sidmouth to Salcombe Mouth along the shingle beach, but the going is tiring.

Great care should be taken to avoid the cliff base because of the frequency of cliff falls.

The cliff comprises an eastward-dipping sequence of Mercia Mudstone, consisting of uniform reddish brown mudstone with some prominent thin beds of green mudstone. Gypsum occurs as nodular geodes and fibrous veins. In the upper cliff, the Mercia Mudstone is overlain in Salcombe Hill Cliff by the Upper Greensand, which is considerably overgrown and not accessible. Offshore, Chapman's Rocks [SY 141 875] beneath Salcombe Hill consist of large blocks of Upper Greensand probably marking the position of an old landslip.

Just east of the mouth of the River Sid, we see a geological repeat of the situation below Peak Hill, for here again the Otter Sandstone can be seen disappearing eastwards beneath the Mercia Mudstone, about 150m east of the river mouth. There are small faults affecting the Otter Sandstone.

Impressive single large joint face in the uniform Mercia Mudstone in Salcombe Hill Cliff.

Mercia Mudstone in Salcombe Hill Cliff, showing how erosion of the cliff has taken place by rock slabs peeling off parallel to the cliff face. Joints at right angles to the cliff have also influenced the process.

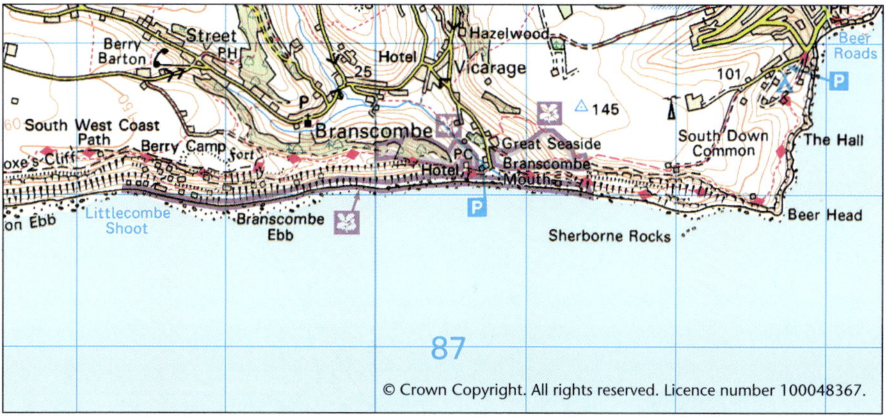

Salcombe Hill Cliff east of Sidmouth, showing rapid erosion in the Otter Sandstone and overlying Mercia Mudstone. A layer of Head is present at the top of the cliff.

The effects of the landslip of January 2006 on the clifftop at Higher Dunscombe Cliff. The slip has eaten into the Clay-with-flints along arcuate fissures.

Higher Dunscombe Cliff showing an apron of debris from higher in the cliff, including large blocks of Upper Greensand material.

The whole stretch of cliff between the mouth of the river and Salcombe Mouth is very unstable, and erosion is proceeding at a rapid rate. Owners of property leading down to Salcombe Hill Cliff have lost many metres of their gardens in recent years and the remains of garden sheds could be seen in 2001 hanging precariously over the clifftop. A considerable length of the coast path east of Sidmouth has now disappeared into the sea, and has been re-routed inland. In March 1999, a rockfall just east of the mouth of the Sid, combined with a high tide, cut off 14 people who had to be picked up by the lifeboat.

Examples of joints can be seen in Salcombe Hill Cliff, and some form impressive single joint planes (page 81). Some joints are parallel to the cliff face, while others are at right angles to it. Erosion working on these joints causes the rock to peel off in large slabs. Faults also influence the way in which the cliff collapses.

At Salcombe Mouth a small fault, not visible, extends along the valley and drops the Mercia Mudstone down to the east by about 15m.

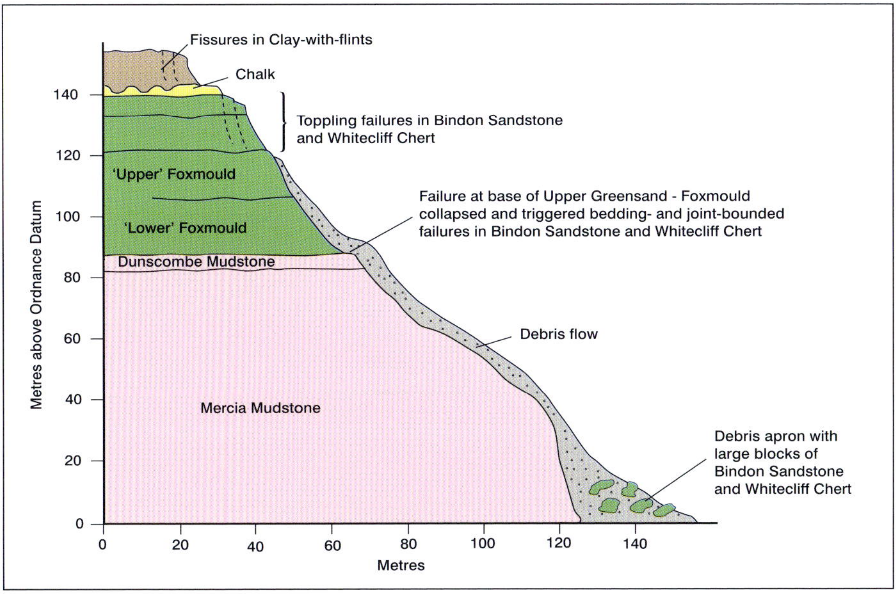

Cross-section showing the strata in Higher Dunscombe Cliff and the effects of the landslip of January 2006.

East of Salcombe Mouth, the lower part of the cliff beneath Higher and Lower Dunscombe cliffs is in Mercia Mudstone, maintaining an easterly dip. Large slips affecting the upper part of Higher Dunscombe Cliff in the early hours of January 4th 2006 deposited large amounts of Upper Greensand material onto the beach, including huge slabs measuring up to 9 x 5 x 2 metres. The total mass of the landslip is estimated to be about 250,000 tonnes. At clifftop level, the width of the collapsed mass was between 5 and 20m. Dr. Ramus Gallois has studied the mechanism of the collapse, and believes that the main failure surface was at or close to the unconformity at the base of the Upper Greensand, probably along a thin bed rich in the clay mineral montmorillonite. Immediately after failure, the Foxmould collapsed and triggered bedding- and joint-bounded failures in the overlying stronger beds (Bindon Sandstone and Whitecliff Chert). The resulting debris flow carried large intact blocks of these units down the cliff in a Foxmould-sand supported matrix, to form a large debris cone covering a 200m – wide stretch of the beach.

Between Salcombe Mouth and Branscombe, fallen blocks of Beer Head Limestone are common on the beach. Ammonites can be found in some of these blocks; for example, on the beach below Higher Dunscombe Cliff they have yielded about 60 ammonites belonging to 12 species.

Hook Ebb consists of large blocks of Upper Greensand, the remnants of old landslips, resting on the foreshore. The nearby cliffs show veins of fibrous gypsum ('satin spar') up to 20cm thick in the Mercia Mudstone. This is the best place on the coast to see samples of orange and cream gypsum with well formed crystals.

Also lying visible on the foreshore are large blocks of cream-coloured Beer Head Limestone. Close inspection shows examples of a type of coral-like bryozoan (*Ceriopora*) which typically occurs near the base of the Beer Head Limestone.

Just east of Hook Ebb, fallen blocks on the beach show examples of sandstones from the Dunscombe Mudstone. The sandstones are very fine grained and some blocks show ripple marks and burrows.

At Little Weston, the small stream along Lincombe descends to the beach. Looking upwards from the beach, the Upper Greensand, now joined by the overlying Chalk, can be seen to form cliffs above the Mercia Mudstone (see below). Pipes of reddish brown Clay-with-flints can be seen in the clifftop.

Cliffs of Upper Greensand overlying Mercia Mudstone above Little Weston.

Coxe's Cliff. Chalk, C, with Clay-with-flints pipes overlies Upper Greensand (Bindon Sandstone, BS and Whitecliff Chert, WC) in the steep part of the cliff. The mainly vegetated lower slopes are underlain by Mercia Mudstone.

From Lincombe eastwards to Weston Mouth, the cliffs of Mercia Mudstone become more vegetated. The green and purple Dunscombe Mudstone can just be made out at the top of the cliff.

From Weston Mouth to Branscombe Mouth is a tiring walk of nearly 4.5km along a shingle beach. The cliffs are in Mercia Mudstone overlain by Upper Greensand and Chalk. A notable feature is the presence of a layer of green and purple mudstone up to about 35m thick that stands out from the mainly red Mercia Mudstone sequence. This Dunscombe Mudstone Formation occurs in the cliff faces below Higher Dunscombe Cliff, but is not easily reached. It is more easily seen in a group of cliffs, degraded steep slopes and slipped masses below Weston Cliff [SY 168 880 to 171 880]. The easterly dip brings the top of the formation down to beach level below Coxe's Cliff. In the lower part of the formation are breccias that may have formed by solution of gypsum or anhydrite. In fresh cliff falls, patches of gypsum can be seen in the breccias – they may pass into thick beds of salt, seen in boreholes farther inland. The formation is notable for the presence of fossils such as *Euestheria* and *Chondrites* and other trace fossils. Sandstones in the formation show ripple marks and burrows.

At Littlecombe Shoot [SY 183 880] there are good views westwards of Coxe's Cliff [SY 179 882], which is composed of Chalk at the top overlying Upper Greensand which in turn overlies red Mercia Mudstone in the lower part of the cliff. Pipes of Clay-with-flints descend into the Chalk at the top of the cliff.

Fallen blocks of Upper Greensand below Coxe's Cliff include examples, probably from the Whitecliff Chert, with layers full of broken shells, mainly oysters. These possibly formed during storms, when the shells were swept into concentrations on the sea bed.

Fossil oyster shell beds in a fallen block of Upper Greensand at Hook Ebb. These may represent storm deposits.

About 400m east of Littlecombe Shoot, and extending eastwards for about 1km is Branscombe Ebb, a foreshore platform covered by boulders of Upper Greensand material which has fallen from the cliffs above. Between the boulders is sand with many colonies of the modern tube-dwelling worm *Sabellaria*.

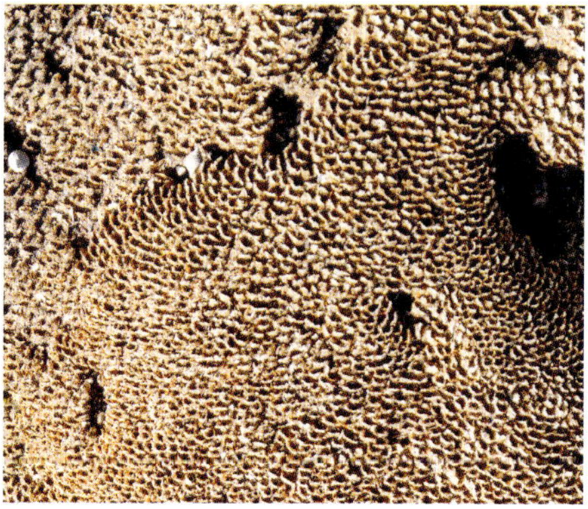

Colonies of the modern tube-dwelling worm *Sabellaria* on the foreshore.

The lower cliffs, made up of Mercia Mudstone, between Littlecombe Shoot and Branscombe Mouth are partly vegetated, and little geology can be seen. Above is a platform partly covered with slipped material, bordered by steep crags of Upper Greensand. Just above beach level below Berry Cliff, a 10m cliff shows gypsum veins interwoven with red mudstone - this is Red Rock (see the photo on page 36), and the Red Rock Gypsum forms a distinctive layer within the Mercia Mudstone of this area.

Between the eastern end of Branscombe Ebb and Branscombe Mouth, the upper part of the beach is flint and chert shingle with some fallen blocks of Cretaceous rocks. The lower part of the beach, seen at low tide, is sandier. However, the make-up of the beach and the thickness of the shingle can vary dramatically at different times of the year.

Clifftop exploration

Taking the clifftop route, climb from Sidmouth up Salcombe Hill Cliff to Salcombe Hill. First cross the small bridge (Alma Bridge) at the mouth of the River Sid. At this point the river hugs closely the cliffs of Otter Sandstone along its eastern side.

A view over the Salcombe Regis Valley towards Higher Dunscombe Cliff, which is made mainly of Upper Greensand capping Mercia Mudstone. The boundary between the formations is evident in the difference in vegetation and topography. A few patches of white Chalk can be seen at the top of the cliff, and a layer of Clay-with-flints forms the plateau top. Sand from a slip in the Upper Greensand has flowed down the cliff onto the beach (this photo was taken before the larger slip of January 2006).

Owing to recent cliff falls, there is no longer a direct route along the clifftop; instead there is an inland detour before the clifftop is regained. The footpath situation is continually changing, so it is a matter of following the official signposts. There is a steady uphill climb along Salcombe Hill Cliff, which is made up of Mercia Mudstone; in places there are glimpses of about 1.2m of stony red clay (Head) in the clifftop, overlying the red mudstone. The base of the Upper Greensand at the eastern end of Salcombe Hill Cliff is easily identified by the change in vegetation to woodland, and the increase in gradient. Follow the coast path steeply up the Upper Greensand slopes. Where the gradient flattens off the Clay-with-flints plateau is reached. From here there are excellent views westwards over Sidmouth and the Sid Valley. Continuing eastwards pass through a gate into a field; there is a cairn with a toposcope here [SY 1382 8765]. Further east pass through another gate [SY 1403 8772] and by the path side is a block of silcrete (see page 26). A few metres farther on is a block of Upper Greensand sandstone called the Frog Stone owing to its resemblance to that amphibian! It was brought to this point in 1964 by Royal Navy helicopter, having been winched from its original location on the Hook Ebb Reef. This is the eastern side of Salcombe Hill. There is a bench and delightful eastward views over the Salcombe Regis Valley towards Higher Dunscombe Cliff.

A panoramic view westwards from Salcombe Hill over Sidmouth and the Sid Valley. Note the flat-topped wooded hills (Peak Hill, Bulverton Hill and Beacon Hill) capped by Upper Greensand and Clay-with-flints. Sidmouth is mostly underlain by Otter Sandstone, with the surrounding farmland occupied by Mercia Mudstone.

Salcombe Hill can also be reached (with less expenditure of effort!) from the National Trust car park [SY 139 882] (with explanatory board), opposite the entrance to the Norman Lockyer Observatory (open to the public at certain times).

From the Frogstone descend the cliff path eastwards. The Upper Greensand is again well defined by steep slopes and vegetation. At the bottom of the valley, the path diverts slightly inland to cross the stream by a small bridge. There is an access point to the shingle beach and the base of the cliffs.

Salcombe Regis valley is a delightful valley about 1km long, closed at its northern end where the village of Salcombe Regis with its Norman church (St Peter's) is sited. There is a car park [SY 148 888] near the church. Several very pleasant circular walks can be planned using the 2½ inch OS map (Explorer 115). Salcombe Regis is the site of old quarries in the Salcombe Regis Stone (see page 110), and the church is largely built of that stone.

Climb steeply eastwards along the coast path from Salcombe Mouth. The cliffs between here and Weston Mouth are Higher Dunscombe Cliff on the west and Lower Dunscombe Cliff on the east, separated by a small combe called Lincombe, around which the coastal path detours inland. The base of the Upper Greensand at Higher Dunscombe Cliff is again easily identified by steeper slopes and change of vegetation. There are exposures of Upper Greensand in the cliff face, where it can be divided into the Foxmould overlain by the Whitecliff Chert which, however, in this area, confusingly, does not contain much chert. At the top of the cliff small patches of white indicate the most westerly outcrops of Chalk along the southern coast of England. As described earlier, there were extensive slips in January 2006 affecting the Upper Greensand and Clay-with-flints in the upper cliffs, and great blocks of Upper Greensand fell onto the beach.

Rempstone Rocks showing Beer Head Limestone (BHL) (with seagull standing on it contemplating the view down into Weston Mouth), overlying Bindon Sandstone (BS), the topmost layer of the Upper Greensand.

At Lower Dunscombe Cliff, just west of Weston Mouth, there is an excellent viewpoint at Rempstone Rocks [SY 162 880]. The Beer Head Limestone can be seen as a distinctive thin bed. Below it lie sandstones (Bindon Sandstone) at the top of the Upper Greensand. Above the Beer Head Limestone is the Chalk which was worked in now overgrown old quarries which extend northwards away from the coast along the west side of Weston Combe towards Dunscombe. This area of old spoil heaps is called Dunscombe Humps. It is uncertain to what extent the sandstones of the topmost Upper Greensand were worked as well as the chalk, for not far away at Dunscombe are the quarries [SY 160 887] which supplied stone for repairs to Exeter Cathedral in the 1970s (page 109).

Now comes a steep descent into Weston Combe, from which there is access to the beach at Weston Mouth if needed. This is followed by a steep ascent from Weston Mouth to the clifftop at Weston Cliff. Just west of Coxe's Cliff, the coastal path swings a few hundred metres inland for about 1km, but comes back to the clifftop at a point above Littlecombe Shoot [SY 183 880]. A steep path from the beach meets the coastal path about 300m west of Berry Camp. Looking down from this point there are crags of Upper Greensand and Chalk forming the top of the cliff. Below these crags is a flattish area which was in former times occupied by small potato fields, but is now the site of a scatter of wooden holiday bungalows.

About 2km west of Branscombe Mouth, the coastal path passes through Berry Camp [centred on SY 188 882], situated above Berry Cliff. Part of the southern side has been lost through cliff erosion. It is a large hilltop enclosure, although it does not occupy an obviously easily defended site. It has not been excavated, but is believed to date from the first millennium BC. The remains of barrows and a possibly prehistoric field system can be seen west of the Camp.

Just east of Berry Camp, the clifftop is extensively disturbed for over 1km as far as West Cliff, above Branscombe Mouth, by old workings in the basal part of the Chalk. The path goes through a succession of humps and hollows which are the spoil heaps from the old workings - this area is called Branscombe Humps. The coast path runs

Branscombe Humps, the remains of old spoil heaps resulting from workings in the lower part of the Chalk.

St Winifred's Church, Branscombe

200-300m inland along the edge of woodland, but better views can be obtained by walking along one of the paths that meander through the Humps. There are glimpses in the clifftop of crags in the topmost sandstones of the Upper Greensand (Bindon Sandstone). Looking northwards, there are occasional views of St Winifred's Church in Branscombe village. There are a couple of linking paths to the village (see Explorer map Sheet 115). From West Cliff there is a steep descent into Branscombe Mouth.

Branscombe Mouth to Seaton (about 4km)

Between Branscombe Mouth and Beer there is a considerable change in the cliffs brought about by the eastward descent of the Upper Greensand and Chalk from the clifftop at Branscombe Mouth to beach level at Hooken Beach. From Hooken Beach, the cliffs are entirely in Cretaceous rocks towards Beer Head and then round the headland all the way to Whitecliff, near Seaton.

This section of our exploration takes us along high Cretaceous (Chalk and Upper Greensand) cliffs, including Beer Head, the most westerly Chalk headland in England. It includes also the spectacular Hooken Landslip. Inland, visit underground workings in the Beer Stone, dating from Roman times (see page 107). Following their formation, the Cretaceous rocks, originally laid down as flat-lying layers or beds, have been bent into a shallow U-shaped fold called a syncline. In such a fold, the layers are inclined inwards towards a central axis. Thus at Whitecliff the beds are inclined towards the west, while the comparable layers in the cliffs at Beer Beach are inclined to the east. The Beer Syncline can be seen on the cross-section at the back of this book.

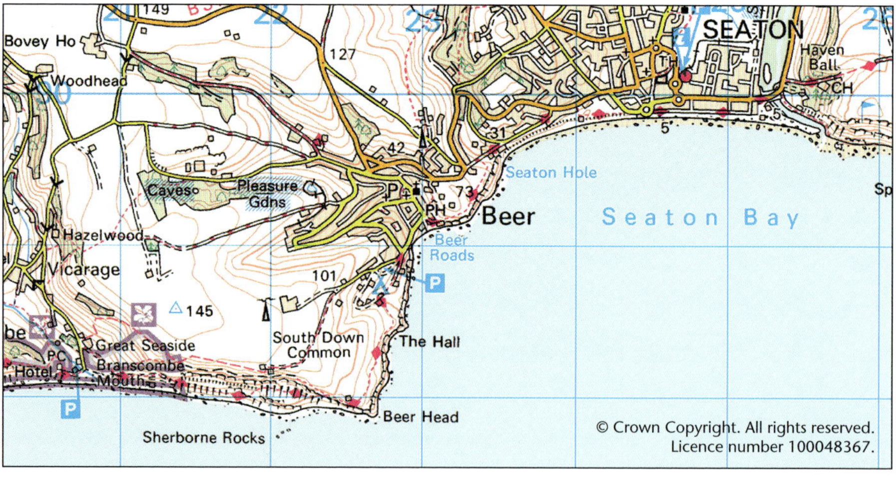

The Hooken Landslip. Masses of Chalk (C) and Upper Greensand (UGS) have slipped down. An adit (A) is visible near the base of the Chalk. In the middle distance is Branscombe Mouth (BM) and in the far distance Sidmouth.

Access and walks

This section is easily accessible along the coastal footpath that extends along the clifftop from Branscombe to Beer, and from Beer towards Seaton. About 500m west of Beer Head a path descends the cliff into the undercliff formed by the Hooken Landslip. There are car parks at Branscombe Mouth [SY 207 882] and at Beer [SY 228 888]. Boats can be hired during the season from Beer Beach, giving the opportunity to examine the geology of the cliffs from a different perspective.

Detail of a wall at St Winifred's Church, Branscombe, showing sandstone from the Upper Greensand and minor chert, C (brown).

Branscombe village

The charming village of Branscombe ('Brannoc's combe') lies about 0.8km inland. Its buildings are scattered along a winding valley leading to the sea at Branscombe Mouth, with two main concentrations, one around the church and another around the Mason's Arms. The church (St Winifred's, after a North Welsh saint who died about 650) is of great interest and shows a process of continuous development from the 11th to the 16th Century.

Near Branscombe Mouth, Great Seaside [SY 2075 8834] is a good example of an Elizabethan farmhouse.

Branscombe was formerly renowned for its early potato crops, and small well-sheltered fields or 'platts' were worked on the cliffs, for example above Littlecombe Shoot. Seaweed collected from the beach at Branscombe Mouth was used as fertilizer and carried to the potato plots by donkey.

At Branscombe Mouth there is a large car park, a café and shop (the 'Sea Shanty') on the site of a former gypsum milling works (see page 114). There is a shingle beach, mainly made up of flints, which at one time were removed for industrial use (see page 115).

Beach exploration - Branscombe Mouth to Beer Head

At the right state of the tide, the geology can be explored from beach level by walking eastwards from Branscombe Mouth along the shingle to Beer Head (about 2km – a tiring walk). **Take care not to approach the cliffs too closely**. Along the way skirt the seaward side of the Hooken Landslip with good views of The Pinnacles. Just offshore from Under Hooken are Sherborne Rocks [SY 221 878], probably the remnants of slipped blocks. The name indicates a link to Sherborne Abbey, which once owned land around Beer. Beyond the slipped material, the high cliffs extending to Beer Head are made up of Upper Greensand and Chalk. The basal part of the Chalk (the Beer Head Limestone) can be seen. It is thicker than at Beer Beach, and a division into three layers is recognisable.

Clifftop exploration - Branscombe Mouth to Beer Head

Eastwards from Branscombe Mouth, there is a steep ascent up the coastal path to East Cliff. Cliffs of Upper Greensand can be seen from the path. About 300m east of Branscombe Mouth there is a choice of paths, which are signposted. The higher path continues upwards to the top of East Cliff; the lower passes firstly through an area of wooden chalets lying below the Upper Greensand cliffs, and then winds for about 1.5km through a fascinating area of tumbled rock, formed by landslipping. The way is rough and not suitable for all walkers, but it is well worth the effort. The path regains the clifftop path by steep steps up the cliff.

About 500m west of Beer Head, there is a magnificent view westwards over one of the most striking landscapes of our area – the Hooken Landslip (see photo on page 91). The last great landslip took place over 200 years ago - in March 1790 - when a

View from beach level of The Pinnacles in the Hooken Landslip. Cliffs around Sidmouth in the background.

great crack opened up overnight in the clifftop, and huge blocks of Chalk and Upper Greensand 250m and over 1/3 mile long slid down towards the sea by between 60 and 80m. The resulting jumbled mass of broken Upper Greensand and overlying Chalk, now much overgrown, is known as Under Hooken. The huge masses of rock were let down in places almost intact, and on the seaward side of the landslip are two columns of Chalk called The Pinnacles. Also visible on the photograph on page 91 is an opening in the cliff face (A) which is often said to connect with the Beer Quarry Caves. In fact the opening comes to a dead end after about 20m.

Continuing eastwards from Hooken Cliffs we reach Beer Head from where, on a clear day there are wonderful views eastwards to the Dorset coast and the Isle of Portland. Beneath the cliff are two precarious-looking columns of white chalk with flints (Upper Chalk). Beer Head is a well-known locality for the Neolithic industry of flint working (see page 104). On top of the cliff are the remains of small roughly square fields with surrounding banks which are similar to the well-known 'Celtic' fields of Wessex.

Beer Head to Beer Beach

The vertical cliffs extend for 1.3km from Beer Head towards Beer Beach, and consist of white chalk overlying Upper Greensand at the base. Depending on the tide, the base of the Chalk and the top of the Upper Greensand can be seen at the base of the cliffs at Beer Beach. At the base of the Chalk at Beer Beach is a distinctive thin (about 0.6m) bed of limestone called the Beer Head Limestone. It is the equivalent of the Lower Chalk elsewhere in England. It rests sharply on the yellowish brown - weathering Upper Greensand (see photos on page 42). The Upper Greensand forms a ledge ('Big Ledge') that extends northwards from The Hall to Beer Beach.

An interesting feature of the cliffs is the influence of marl bands on the solution of the Chalk and its effect on stability (see pages 44-45)

Beer village

The charming village of Beer is sited along a valley formed along the axis of the Beer Syncline. The name derives from the Old English word *'beare'* meaning grove. A stream (Beer Brook) runs beside the main street down to a beach of flint shingle sheltering beneath white Chalk cliffs. The village was formerly a centre for quarrying, and the famous Beer Stone quarries are about 1.6km west of the village (page 107). There are good examples of the use of Beer Stone, sometimes in conjunction with flint, in the village buildings. The village was also celebrated for its lace. Fishing is another occupation that continues to the present day. The celebrated smuggler Jack Rattenbury, born in 1778, lived in Beer and still remains a local folk hero!

From Beer, the coast path extends to Seaton over the clifftop. On the way there are superb views looking back to Beer Beach. The path used to pass beneath the prominent rock face of Annis' Knob, which consists of white Upper Chalk with flint layers, gently inclined to the west, but cliff falls in 2001 have led to re-routing of the path. The cliffs below Annis' Knob are in Middle Chalk.

Seaton Hole and Whitecliff

At Seaton Hole [SY 235 896], a fault separates pale grey and white Cretaceous rocks on the west from red Triassic mudstone on the east, providing a dramatic colour contrast. Triassic rocks were last seen at Branscombe Mouth, but in the intervening ground they are concealed beneath the later Cretaceous rocks in the Beer Syncline. Along the Seaton Hole Fault, the Cretaceous rocks have dropped downwards to the west by about 60m relative to the Triassic rocks on the east. The actual line of the fault is covered by slipped material down which the path runs to the beach.

Just west of Seaton Hole, Whitecliff [SY 2344 8942] reveals almost the whole of the Upper Greensand and some of the overlying Chalk. Although the cliff is largely inaccessible, the subdivisions of the Upper Greensand can be seen from a distance. The lowest subdivision is the Foxmould, about 10m thick; it consists of yellowish brown weathering sandstones. The base is not seen at Whitecliff. The overlying subdivision is the Whitecliff Chert (formerly called the 'Chert Beds'), about 15m of pale sandstones with irregular layers of chert. The topmost subdivision of the Upper Greensand is the 5m-thick Bindon Sandstone (formerly called 'Top Sandstones'). To avoid the dangerous cliffs, the various subdivisions of the Upper Greensand can be examined in fallen blocks, in some of which are fossil shells, including remains of oysters and crustacean burrows.

The Bindon Sandstone is overlain at the top of the cliff by the white Chalk at the base of which is a thin layer of Beer Head Limestone. Excellent views of Beer Beach can be obtained by a short careful scramble west over the boulders at the base of Whitecliff.

Whitecliff, between Beer and Seaton. Here the cliffs show Upper Greensand overlain by Chalk. The Upper Greensand is divided into Foxmould at the base, overlain by Whitecliff Chert and Bindon Sandstone.

Seaton to Pinhay Bay (about 8km)

This section of the coast is included in the Axmouth to Lyme Regis Undercliffs National Nature Reserve. It is one of the largest active coastal landslip systems in Western Europe (see photo opposite). The area has recently (2006) been described in a fascinating book '*Exploring the Undercliffs*', by Donald Campbell. Consequently, only an abbreviated description is given here, and the landslips have not been described in any detail. The book contains (on its pages 14 to 15) a photographic panorama of the geological strata from Axmouth to Lyme Regis.

A detailed account of the geology of the cliffs between Seaton and Culverhole Point is given by Dr Ramues Gallois (2003: *Geology walks in and around Dorset,*) (see Further Reading section, page 124).

Access and walks

In Seaton itself there are large car parks by the tramway terminus. For walks along the coast east of the mouth of the River Axe, there is limited parking space beside the B3172 Seaton to Axmouth road [SY 253 901]. Walks along this section, either along the beach or through the landslipped Undercliff, can be quite arduous. There are no permitted paths allowing access from the Coast Path to the beach or vice versa.

! When planning your walk please note that Culverhole Point cannot be traversed at high tides.

The coast is perhaps most easily appreciated from the sea, and boat trips run from Beer and Lyme Regis. The eastern end of the section (Pinhay Bay) can be approached from the Lyme Regis end (about 3km to Pinhay Bay).

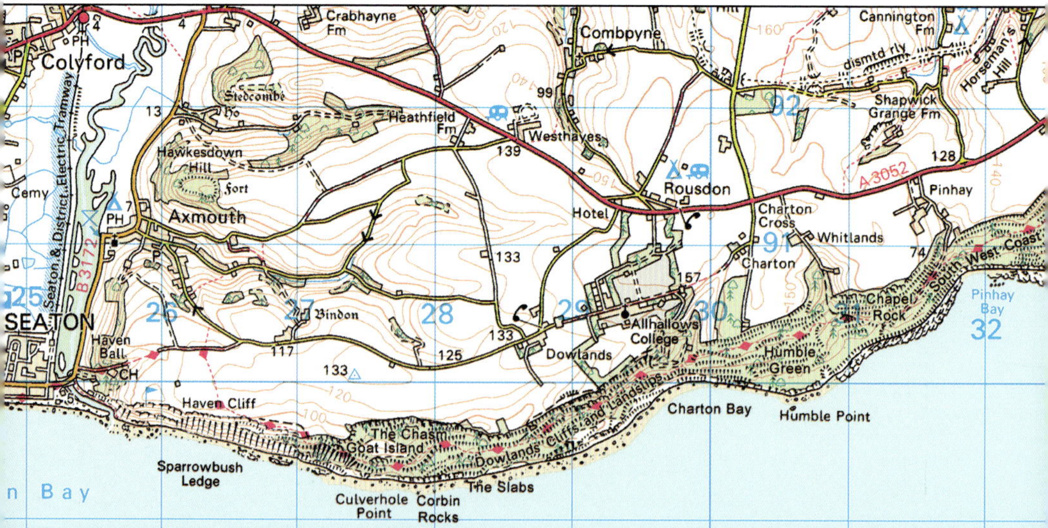

The Bindon Landslip: The Chasm (TC), Goat Island (GI) and Culverhole Point (CP).

Seaton

Seaton is a pleasant resort at the mouth of the River Axe. There is an interesting small museum. Along the seafront between Seaton Hole and Seaton, there are low cliffs of Mercia Mudstone with several small faults. A shingle ridge extends along the coast and nearly cuts off the mouth of the River Axe – a situation comparable to that at Budleigh Salterton.

Seaton to Pinhay Bay

The cliffs along this, the final section of our journey along the coast, show the highest part of the red Mercia Mudstone, the grey Blue Anchor Formation, and the Jurassic Blue Lias and Charmouth Mudstone. In addition, there are excellent sections in slipped Upper Greensand. The upper parts of the cliffs are heavily landslipped and include the famous Bindon Landslip.

The walk along the base of the cliff begins at the mouth of the River Axe. The lower cliff from the river mouth for about 1.6 km eastwards consists of red Mercia Mudstone towards the western end, with grey or green mudstones of the Blue Anchor Formation towards the eastern end. At the western end, below Haven Cliff, the Mercia Mudstone is folded into a well-developed small anticline.

The anticline below Haven Cliff.

The upper cliff (Haven Cliff) consists of Upper Greensand and Chalk, separated from the lower Mercia Mudstone cliffs by a bench on which are collapsed masses of the Cretaceous rocks. Hollows and pipes within the Chalk at the top of the cliff are filled with orange-brown Clay-with-flints.

About 500m west of Culverhole Point, exposures [SY 273 894] on the foreshore show a low tipped mass of 'White Lias') (Langport Member of the Lilstock Formation of the Penarth Group) consisting of white-weathering limestones. On the foreshore in front of the 'White Lias', black mudstones of the Westbury Formation (with the distinctive bivalve *Rhaetavicula contorta*) can be seen, as well as the overlying pale green mudstones of the Cotham Member. At the base of the Westbury Formation, but not normally exposed, is the famous 'Rhaetic Bone Bed' which contains fish and marine reptile teeth and bones.

'Cowstones' on the beach east of Culverhole Point.

Culverhole Point. Chalk at the top of the cliff; Upper Greensand below. The lower part is mainly Whitecliff Chert, with Foxmould right at the base of the cliff.

Continuing eastwards towards Culverhole Point, the beach is strewn with rounded boulders of calcareous sandstone which are derived from the lower part of the Upper Greensand (the Foxmould) as well as larger blocks from the higher part of the Upper Greensand (Whitecliff Chert and Bindon Sandstone). The boulders have been referred to as 'Cowstones' owing to their supposed resemblance when seen on the foreshore to a herd of resting cows!

Culverhole Point itself is an excellent place to see almost the whole of the Upper Greensand and part of the overlying Chalk in a large mass which has slid down almost intact from higher in the cliff. The lower half of the cliff consists of Whitecliff Chert – yellowish brown calcarenite with layers of black chert – overlying unweathered grey Foxmould at the base of the cliff. The junction of the Whitecliff Chert with the

'White Lias' near Culverhole Point

Foxmould is marked by two erosion surfaces, or hardgrounds.

Between Culverhole Point and Corbin Rocks, there are cliff sections in up to 8m of Whitecliff Chert (Upper Greensand).

Continuing eastwards, there is a distinctive area of gently dipping layers of limestone belonging to the Jurassic Blue Lias, and known as 'The Slabs' [SY 286 894]. The upper surface of the most prominent limestone bed shows large examples of the early Jurassic ammonite *Coroniceras* which identifies the limestone bed as the Grey Ledge bed of the Blue Lias. Almost complete specimens of marine reptiles have also been found at this site.

West of Charton Bay, the foreshore is in 'White Lias', and behind there are quite prominent cliffs in landslipped material consisting of chalk fragments. The foreshore is bouldery. There are the remains of a wrecked boat offshore, and remnants of a small 'harbour' at the west side of Charton Bay.

The Slabs, inclined layers of limestone and dark grey mudstones belonging to the Jurassic Blue Lias Formation.

Examples of the ammonite *Coroniceras* on bedding surfaces at The Slabs.

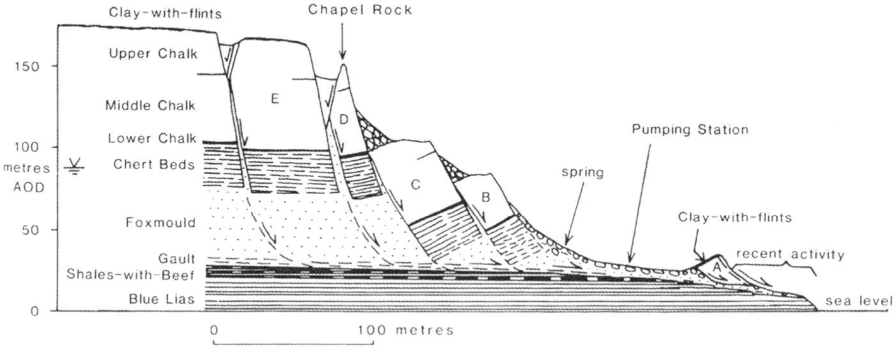

A cross section through the cliffs below Chapel Rock showing the structure of the landslips. Chert Beds = Whitecliff Chert.

The cliff on the west side of Charton Bay consists of about 29m of grey-green mudstones with some paler bands, of the Blue Anchor Formation; the base is concealed beneath the beach deposits. Just west of the cliff, a fault brings in Blue Lias in tilted blocks.

Between Humble Point (east of Charton Bay) and Pinhay Bay, there are low cliffs of landslipped material. Halfway along, up in the cliffs we can see the white Chalk pinnacle of Chapel Rock. A large spring issues from below Chapel Rock and is the water supply for Lyme Regis. Threats to the water supply by landslips led to a study by Peter Grainger. He showed that the cliff above the spring was made up of four large blocks of Chalk and Upper Greensand, thought to have slid forward along the Gault. During this sliding, the fine sand of the Foxmould had liquefied, allowing the blocks to subside and rotate. A cross section is shown on the previous page.

From Charton Bay, shingle, mostly of Upper Greensand chert and flint, extends round to Humble Point. Loose blocks of Beer Head Limestone lying on the foreshore at Humble Point have yielded ammonites. The area of open ground above Humble Rocks is called Humble Green. At Humble Point, there are reefs of 'White Lias', and Blue Lias reefs appear on the foreshore farther east.

The appearance of the ammonite *Psiloceras* near the base of the Blue Lias marks the start of the Jurassic period.

Pinhay Bay marks the easternmost extent of our journey along the coast. Here we are only about 3km west of Lyme Regis, and the cliffs east of Pinhay Bay are best reached from that town. The cliffs at Pinhay Bay (page 102) show the whole of the Blue Lias (about 24m thick), with the lower part of the Charmouth Mudstone (Shales-with-Beef) at the top. Fossils to be found in the Jurassic rocks here include the bivalves *Plagiostoma giganteum* and *Gryphaea arcuata*, examples of the nautiloid *Cenoceras*, and very large specimens of the ammonite *Coroniceras conybeari*.

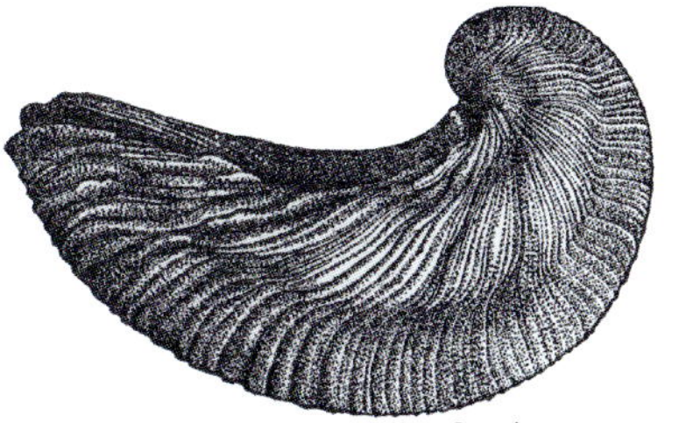

Gryphaea arcuata.

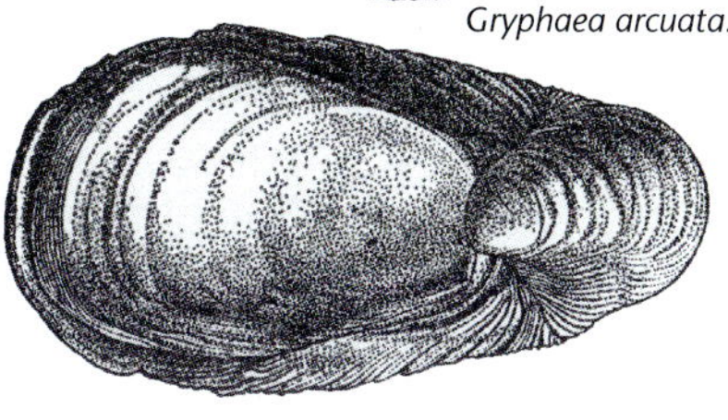

The boundary between the Triassic and Jurassic periods occurs within the lower part of the Blue Lias. The start of the Jurassic is marked by the first appearance of the ammonite *Psiloceras* about 2.5m above the base of the Blue Lias.

At the base of the cliff are about 9m of Penarth Group (Langport Member or 'White Lias') consisting of limestones with slumps and channels.

A return to the starting point can be made by walking back along the base of the cliff (having checked the tide so that Culverhole Point can be traversed).

Pinhay Bay. Most of the cliff is in Blue Lias (grey mudstones with thin limestone beds). At the base is pale grey 'White Lias'.

USEFUL STONE AND MINERALS

Gypsum vein.

Since prehistoric times the area has provided useful natural stone and minerals to sustain the economy of the local society. A surprising variety of natural materials has been won from the ground. Among these are included flints, building stones, lime and marl, brick and tile clays, gypsum and salt.

Flint

Prehistoric flint working at Beer Head

The flints that occur in irregular lines within the Chalk of Beer Head were used by Neolithic Man for the manufacture of a whole 'toolkit' of implements such as leaf-shaped arrowheads, axes, scrapers, piercers, borers and knives. One of the best quality flint bands in Europe occurs near the top of the cliff at Beer Head and seems likely to have been the source of much of the flint. It consists of excellent quality jet-black flint. Aileen Fox notes that only small implements were made at Beer Head and rough-outs for hand axes are significantly absent. The amount of waste from flint working at, for example, Hembury Fort suggests that Beer Head was the source of the raw material and not a place where the finished products were made. People looking for flint from Beer Head probably made regular visits either by sea, landing at Beer Beach, or travelled overland by ridgeway tracks.

Collections of Neolithic flint implements, many of them beautifully worked, have been made from surface material at Mutter's Moor and from Salcombe Hill. A display at Sidmouth Museum shows flint implements, especially from Mutter's Moor, Sidbury Castle and various other localities around Sidmouth. Examples of Iron Age flint sling-stones found at the entrance to Sidbury Castle hillfort can also be seen.

Manufacture of gunflints at Beer Head

Many thousands of years later, in the 17th to 19th centuries, the flints of Beer Head found another use as gunflints in flintlock firearms. In 1879 Skertchley gave a detailed account of gunflint workings and production in England, especially at Brandon in Norfolk. He also described the industry at Beer from an account by the well-known Sidmouth antiquarian P O Hutchinson, who reported that the locality where gunflints were worked was the undercliff [Hooken Landslip] on the west side of Beer Head. He says :

"This undercliff was one of the places to which the gun-flint makers used to resort in order to follow their calling, as the landslip had probably disinterred plenty of black flints ready for use, or made them more accessible than before. Heaps of flint chips and splinters marked the spots where they had laboured...The men would go out of a morning with their knapping hammers from Beer and Branscombe and there work till evening. I never heard how many they would make in a day. The trade diminished after the Battle of Waterloo, and ceased on the introduction of the percussion cap."

Skertchley continues: " *The chips are now all dug into the ground, and the locality (which is just below the words "Signal Staff" on the Ordnance Map)* [about SY 219 882] *is fast losing all traces of the old trade.*" Since this was the situation in 1879 there is, as might be expected, now no evidence on the ground of this interesting local industry.

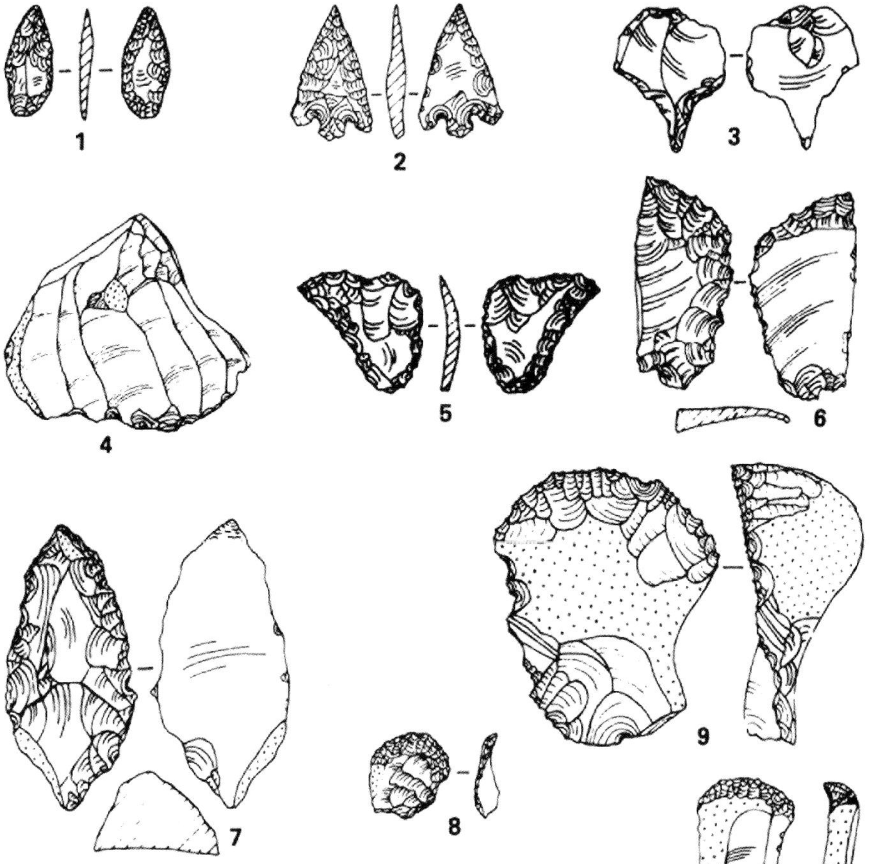

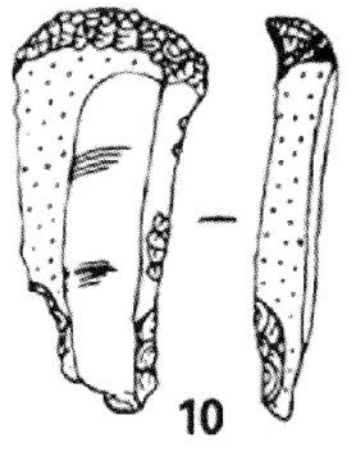

NAMING OF THE FLINTS

1 Neolithic leaf arrowheads.
2 Barbed and tanged arrowhead, Beer flint.
3 Piercers and borers.
4 Typical Neolithic core with long narrow flakes struck off.
5 Transverse late Neolithic arrowheads, *petit tranchet* derivatives.
6 Single-edged knife, Beer flint.
7 Fabricators; thick heavy-duty tools, all with much wear.
8 A typical "button" scraper.
9 & 10 Neolithic flake scrapers, typically with steep scraping angles.

By Sheila Pollard & Sheila Luxton
Devon Archaeological Society

Stone for building

A variety of geological formations have been used for building stone in structures ranging from cathedrals to humble garden walls. Apart from quarried stone, pebbles of flint (from the Chalk), chert (from the Upper Greensand) and quartzite (from the Budleigh Salterton Pebble Beds) have been picked from soils or from beaches to use in buildings, walls, and cobbled floors and yards.

Beer Stone

Probably the best known building stone in the area is the celebrated Beer Stone, noted for its use in many prestigious buildings, especially Exeter Cathedral. It is of such high quality as a freestone that it has been worked since Roman times from quarries and mines about 1.6km west of Beer. The oldest workings [SY 215 894] are to the south of the lane and are called the Old Quarry or Beer Quarry Caves. The 'New Quarry' workings [SY 215 898] lie to the north of the lane.

The Beer Quarry Caves are open to the public and are well worth a visit. The workings are entirely underground. The earliest underground workings at Beer Quarry Caves show the characteristic rounded arches found in other Roman workings. The square Saxon workings at Beer Quarry Caves are in marked contrast to the arched Roman workings. The Roman workings later became linked to give a large chamber. All the post-Roman workings were mainly square in plan and of the `pillar-and-stall' method, where unquarried columns of stone were left at regular intervals to support the roof. This pattern of working continued for the next 900 years, to give the present impressive complex of galleries and caverns that can be seen today.

During the winter the Beer Quarry Caves are closed owing to the fact that they are the home of rare species of pipistrelle bat. A fascinating account of the workings is given in the booklet by John Scott and Gladys Gray (see Further Reading, on page 124).

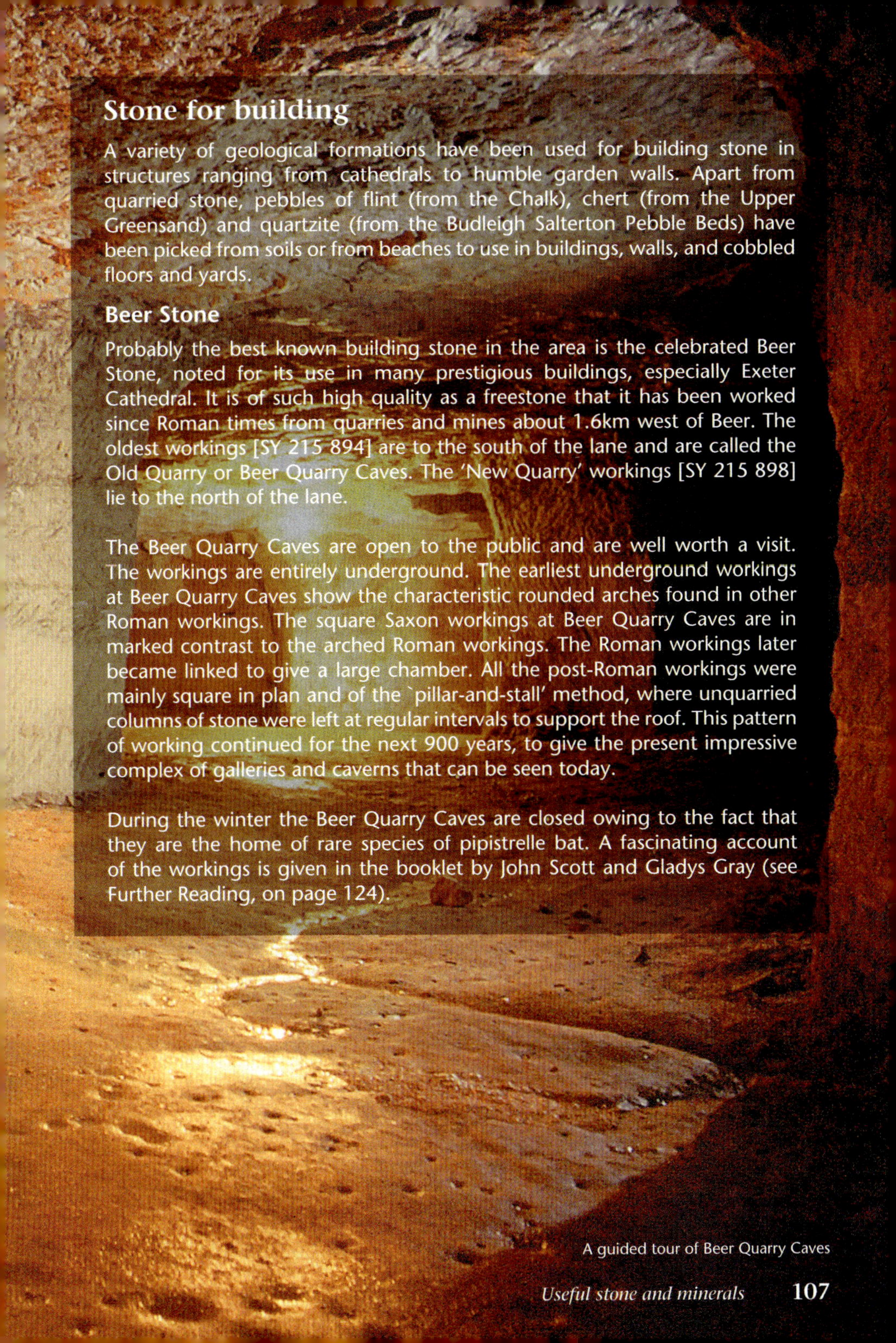

A guided tour of Beer Quarry Caves

It has often been said that the Beer Quarry Cave workings extend to the cliff at Hooken where there is an adit in the cliff face, but in fact there is no connection. De la Beche (1839) noted that a shaft had been sunk on the hill at 'Branscombe Cliff' and a level driven from that shaft, but '*the work does not appear to have been profitable, as it is now abandoned*'. The original location of the shaft is not known. Evidence for some past working of Beer Stone at Hooken is provided by Jukes-Browne who, writing in 1903, noted that near The Pinnacles in Under Hooken, dressed blocks of stone still lie about on the slopes.

The 'New Quarry' workings date from the later part of the 18th Century and were largely underground. The Chalk was worked for lime from an opencast quarry. The quarry is currently disused and is not open to the public. A large modern limekiln is sited in the quarry. Before closure of the quarry in 2003, Beer Stone was extracted on a small scale on demand using large stone saws. There is a very extensive network of underground passages which were apparently used during the Second World War to store munitions.

The quarry cottages lie along the north side of the lane on the way to Beer and bear a stone with the construction date of 1883. Nearby Orchard Cottage is a charming thatched cottage built of Beer Stone.

Orchard Cottage, constructed of Beer Stone.

Geologically, the Beer Stone is an unusual local development within the mainly nodular lower part of the Middle Chalk. It consists of about 4m of creamy-white, smooth-textured, fine-grained homogeneous limestone. In the past it was supposed that it consisted of the comminuted shells of the bivalve *Inoceramus*, but recent study suggests that the Beer Stone also includes much echinoderm debris. In certain lights, the crystalline fragments give a characteristic sparkling appearance to the stone.

The recently quarried stone is relatively soft and easy to work, but hardens on exposure. It is easily carved into complex shapes and hence was favoured for decorative features on churches (particularly window frames) and other important buildings. However in outdoor use it has suffered badly from atmospheric pollution so that, for example, extensive restoration work has had to be carried out on the Beer Stone figures on the image screen on the west front of Exeter Cathedral.

Exeter Cathedral is built largely of Salcombe Regis Stone. Beer Stone has been used for the figures on the image screen on the west front.

Despite the large quantities of stone that were removed by the Romans, there are few definite examples of its use in buildings. Excavation of the Roman Villa at Honeyditches, Seaton, showed it in use in the bath house. In Exeter, pieces of Beer Stone were re-used in the west doorway of St George in South Street. The Saxons used Beer Stone as can be seen by the Saxon Cross in Colyton Church. Beer Stone was used extensively in Exeter Cathedral from the 11th Century and the stone became widely used throughout southern England in the 13th to 15th centuries; it has been used in just about every church in East Devon. Other structures that have incorporated Beer Stone are Westminster Abbey, the Tower of London, London Bridge and Winchester Cathedral. After about 1540, for the next 300 years, the stone was only used for secular buildings. The Victorian religious revival led to a renewed demand for Beer Stone in church restoration and construction and the quarries became very busy. In 1883 the Bath Stone Co. took over and began development of the 'New Quarry' north of the lane. This was because the working face of the old workings was now 400m from the entrance and, although the bed was known to extend to the cliff face at Hooken, it was affected by a geological fault. Another point in favour of the northern quarry was that it was in sunshine for much of the day, so speeding up the drying-out of the stone.

Beer Stone has been used extensively in Beer village, and in some places there is effective use of a combination of Beer Stone and flint, for example in Diamond House, Fore Street. Also in Beer, attractive art pieces of carved Beer Stone can sometimes be seen on display in the Marine House Gallery, Fore Street.

Salcombe Regis Stone

Salcombe Regis Stone is a honey-coloured calcareous sandstone occurring within the uppermost part of the Upper Greensand (Bindon Sandstone). Its main claim to fame is that it was used for the construction of much of the visible fabric of Exeter Cathedral. It has also been used for the construction of Salcombe Regis Church, Sidmouth Church, Sidbury Church and Woolbrook Church. The stone was also used for secular buildings in the vicinity of Salcombe Regis, for example the barns at Slade Farm and Slade House itself. It has been worked from at least four quarries in the Salcombe Regis and Dunscombe areas: North Combe Quarry and South Combe Quarry in the Salcombe Regis Valley; Dunscombe Manor Quarry; and Lincombe Quarry.

North Combe Quarry [SY 1487 8872]

This quarry, sited on the hillside about 130m SSE of Salcombe Regis Church, was the largest of the Salcombe Regis Stone quarries. It is now completely overgrown and there is no sign of any stone faces. It is believed to have been the source of the bulk of the stone for Exeter Cathedral, probably from the 12th Century onwards, and for Salcombe Regis Church itself. Records show that stone for the cathedral was carted down the Salcombe Valley to the cliff edge, let down on ropes or sleds to the beach, and loaded on barges for transport to Exeter.

South Combe Quarries [SY 1431 8808; SY 1435 8809]

There are two small quarries on the west side of the Salcombe Regis Valley, just north of South Combe Farm. Their size indicates that little stone was extracted, possibly because the quality was poor. Hutchinson records a visit on January 9th 1859 to a new quarry which had been opened on the west side of the Salcombe valley, about halfway between Salcombe Church and the sea, above South Combe Farm. This seems likely to be either the quarry, remains of which are still visible [SY 1431 8808] or another nearby [SY 1435 8809]. He records that the stone for re-building some parts of Sidmouth parish church was to be brought from this quarry. However he noted that some of the blocks he had seen were so soft and friable that he was doubtful about its value as a building stone. The situation is complicated by the fact that there is an old limekiln [SY 1432 8808] near one of the quarries, so that some of the stone may have been dug for lime as well as building stone.

The tower of Sidmouth Church constructed of honey-coloured Salcombe Regis Stone.

Detail of a wall at Sidmouth Church showing the use of Salcombe Regis Stone. Note the variable resistance of the stone to weathering.

Dunscombe Manor Quarry

Stone from a quarry [SY 1601 8874] at Dunscombe Manor was used for the reconstruction of Sidbury Church about 1445, and for Sidmouth Church. It is also possible that it provided stone for Exeter Cathedral in the medieval period. It was re-opened in 1928 for the building of the modern church at Woolbrook in Sidmouth, and again in the 1970s to obtain stone for repairs to the south tower of Exeter Cathedral. An interesting account of the search for a source of stone for the cathedral is given by Deryck Laming (see Further Reading section, on page 125).

The quarry lies within a caravan site and is sited within the Bindon Sandstone division of the Upper Greensand. Deryck Laming recorded Beer Head Limestone overlying the Salcombe Regis Stone beds, and there is a disused pit in the Middle Chalk on the hillside above the quarry.

South of the Dunscombe Manor Quarry, old Ordnance Survey maps show extensive old workings adjacent to Dunscombe Coppice, along the west side of Weston Combe between Dunscombe Manor and Lower Dunscombe Cliff. These workings were for chalk as a source of lime. It is uncertain whether any Salcombe Regis Stone was also dug. One of the limekilns is still reasonably well preserved.

Lincombe Quarry [SY 1564 8818]

There is another small disused quarry at Lincombe which seems to have been at some time a source of Salcombe Regis Stone.

Stone from the Otter Sandstone

The Otter Sandstone is mostly rather weakly cemented, but the better cemented layers of conglomerate resist weathering better and have been used in buildings in the Sidmouth and Budleigh Salterton areas. Many walls in Budleigh Salterton and Sidmouth are constructed of Otter Sandstone conglomerate.

Use of conglomerate from the Otter Sandstone in a wall in Sidmouth.

Budleigh pebbles

The well rounded quartzite pebbles and cobbles from the Budleigh Salterton Pebble Beds - locally known as 'popples' or 'Budleigh Buns' – occur not only on the actual outcrop of the pebble beds, but also scattered in great numbers in the soils of adjacent areas. They have been widely used for walls, floors and, occasionally, whole buildings.

Close-up view showing use of flint in a wall at Manor Road, Sidmouth.

Near the sea-front at Salting Hill, Budleigh Salterton, a wall [SY 0714 8198] is constructed of evenly-sized cobbles of quartzite from the Budleigh Salterton Pebble Beds, with an upper course of irregular blocks of conglomerate and some red sandstone from the Otter Sandstone (see photo, page 67).

Flint, chert and silcrete

Flints have been widely used in the Sidmouth area to build walls and to face buildings. The angular flint, chert and silcrete pebbles were mainly collected from stony soils, while the rounded types were probably collected from beaches. There are many attractive flint walls, constructed of irregular unworn flints, in Sidmouth, for example in Manor Road (see the photo on the previous page). Pebble Cottage [SY 1231 8748] at the junction of Seafield Road and Station Road in Sidmouth has walls of rounded flints. There are interesting examples of brick and flint cottages at Sea View, Beer.

Pebble Cottage, Sidmouth, showing the use of rounded flint pebbles in the construction of the wall.

Cob

Clayey formations of the area were used in the past as cob (a mixture of clay and straw) to construct walls and buildings. Some of the many pits on the Mercia Mudstone may have provided clay for this purpose. They would tend to be close to the site of the building to avoid having to move the heavy clay very far.

Cottages at Sea View, Beer, showing the use of flint and brick in combination.

Lime and marl

To produce agricultural lime, limestone is burnt in kilns. The lime is spread on fields to improve ('sweeten') soils. The limestone was imported from quarries in the Devonian limestones of south Devon, and was also produced from local quarries in the Chalk and to a much lesser extent, the Upper Greensand. The imported limestones were brought to our area by boat, unloaded onto the shingle beaches and then taken to the kilns. These kilns were mostly sited close to the shore to avoid having to move the heavy limestone too far. The Limekiln car park at Budleigh Salterton is named from the former limekilns sited there. The remains of the formerly

Old limekilns at Limekiln car park, Budleigh Salterton.

6m-high kilns with bricked-up openings can be seen on the island between two lanes by the car park at the bottom of Saltings Hill. Limestone of Devonian age from Brixham was tipped into the sea at high tide, and thence loaded into wagons on a tramway which ran to the kilns. For many years there was a limekiln on the clifftop at Jacob's Ladder (now the site of Connaught Gardens). Other limekilns are known from Ladram Bay. Coal was also needed as fuel for the kilns and must also have been brought in by boat.

Limestone for the production of agricultural lime was also extensively dug locally from the Chalk, especially in the Branscombe area where the spoil heaps ('Branscombe Humps') testify to the large extent of the workings. There are no obvious limekilns in this area. The quantities involved are such that it must have been an important local industry over a long period. Chalk from the large modern quarry (New Quarry) [SY 215 898] opposite the entrance to the Beer Quarry Caves was burnt for lime in a large modern kiln.

Old kilns in association with Upper Greensand quarries suggest that the calcareous sandstones of that formation were also burnt for lime. Old quarries and a limekiln [SY 1432 8808] are sited about 100m north of South Combe Farm on the west side of the Salcombe Regis Valley.

Marl is the name given to mudstone with a calcareous content. There was a long tradition in Devon of digging pits into the mudstone of the Mercia Mudstone and Aylesbeare Mudstone to obtain marl for spreading on the land to improve soils. There are many small pits scattered across the outcrop of the Mercia Mudstone and Aylesbeare Mudstone. The map on page 56 shows the distribution of these marlpits in the Sid Valley.

Bricks and tiles

Old Ordnance Survey maps show a brick pit [SY 120 887] and kiln on the west side of the road opposite the former Sidmouth railway station, and another [SY 1220 8875] just east of the station. A 'Brick and Tile Works' with an associated pit centred on [SY 1320 8955], in the general area of Byeside Road and Hides Road, is marked on the old Ordnance Survey map about 500m SSW of Sidford.

Gypsum

Gypsum was worked in the 19th Century from the Mercia Mudstone at Branscombe, about 200m east of Littlecombe Shoot. Vancouver, writing in 1808, noted that the gypsum used to be sold '*at 2s. 6d. per hundredweight*'. The Littlecombe Shoot mine is not marked on the 1882 edition of the 1:2500-scale Ordnance Survey map, but the area is subject to frequent landslips so that any old workings might be rapidly covered. A grinding mill, powered by a mill wheel driven by the small stream which flows along the valley floor, was built at Branscombe Mouth to process the gypsum, and operated from 1855 to the 1890s. It was built by Mr Wheaton of Exeter, and was shown in work in a painting by Sir Kevin Spencer, dated about 1870. The remains of the mill wheel were dismantled in the 1960s. The mill leat runs on the uphill side of the beach café, and the stone foundations are still visible where the leat meets the low sea cliff.

Salt

Salt has been extracted from saltmarshes along the East Devon coast since at least Domesday times. The town of Budleigh Salterton is named from the salt pans (salterns) at the mouth of the River Otter. The village of Otterton, about 1.6km north of Budleigh, was a centre for salt-workers, and in the 11th Century Domesday noted thirty-three. In the Branscombe area, some field names include the word 'cellar', which may be derived from the Latin 'salaria'.

Owing to the flooding of the estuary over many years, all signs of the industry have been removed at Budleigh Salterton, although at Seaton remains of at least twenty salt pans are reported. The extraction methods were probably similar to those used elsewhere around the coast of Britain. Salt, mixed with the mud and sand of the estuary, was collected into mounds and then evaporated out in large hoppers. It was then heated in shallow containers until it crystallized. The fuel used in East Devon must have been wood. By the 17th Century, the efficiency of the German salt mines had made this laborious process uneconomic.

Beach pebbles

Beach pebbles – mainly flints, and quartzites from the Budleigh Salterton Pebble Beds have been widely used in buildings and walls and particularly for cobbling.

The beach pebbles from Branscombe Mouth (and also from Seaton) were at one time removed by a company called "Pebbles and Flints" which supplied companies such as ICI and Colgate with stones to use as grinding pebbles because they were so hard. The pebbles were hand selected by 'pebble pickers'. Part of the Post Office Tower in London is faced in black flints from East Devon. Removal of pebbles stopped in the 1970s after concern about the amount of material being removed, and its affect on the beach profile.

In Victorian times, beach pebbles of silcrete (see page 46) and other stones were polished and sold to visitors as 'Sidmouth pebbles'. Hutchinson, in his '*History of Sidmouth*' notes that the main types of Sidmouth pebbles were chalcedony, agate, onyx and sardonyx (not common at Sidmouth), jasper (found on Sidmouth beach, but mostly from Budleigh Salterton beach), conglomerate [silcrete], and manganese (so-called) – often with splinters of flint in it. Hutchinson recorded that the common price of cutting an agate, chalcedony or other siliceous pebble, using diamond dust, was 5½d per square inch. For cutting and polishing the price was 7½d per square inch. In Sidmouth Museum an octagonal Victorian table set with pebbles from Sidmouth beach, cut and polished by the stone mason James Newman, is on display.

GLOSSARY

Adit. A horizontal entrance or passage to an underground mine.

Alluvium. Deposits of clay, silt, sand, gravel and peat laid down on the floodplains of modern rivers.

Ammonite(s). A general term for a group of coiled fossils found especially in the Jurassic.

Amphibians. Animals living both on land and in water.

Angular unconformity. A type of unconformity occurring between two units of rocks in which the underlying older rocks dip at a steeper angle than the overlying younger strata.

Anticline. A fold, the centre of which contains older rocks, and which is convex upwards.

Anhydrite. A mineral consisting of anhydrous calcium sulphate. It represents gypsum without the water of crystallisation, and alters easily to that mineral.

Arthropod(s). A member of the group of invertebrate animals belonging to the phylum Arthropoda, characterised by jointed appendages and segmented bodies. Includes trilobites, crustaceans etc.

Aylesbeare Mudstone. A group of Permian strata underlying the Budleigh Salterton Pebble Beds between Budleigh Salterton and Exmouth and consisting mainly of reddish brown mudstones with some sandstones.

Baryte. A mineral, barium sulphate, recorded in nodules within the Mercia Mudstone.

Bed(s). A layer in a sedimentary rock of generally uniform composition, separated from rocks above and below by more or less well defined boundaries.

Bedding. The arrangement of sedimentary rocks in beds or layers of different thickness and character.

Beer Head Limestone. The basal unit of the Chalk in our area, it is the equivalent of the Lower Chalk elsewhere in England. It consists of dense, richly fossiliferous limestone.

Beer Stone. An unusual local development of limestone about 4m thick within the lower part of the Middle Chalk between Beer and Wilmington. It is the only true freestone in the Chalk of Britain and has been mined west of Beer.

Beer Syncline. A bowl-shaped fold that affects the Upper Greensand and Chalk between Beer Head and Seaton Hole.

Bindon Sandstone. The highest (youngest) of the three subdivisions of the Upper Greensand in our area. It consists of up to 8m of fine- to coarse-grained sandstones and calcarenites. Some chert layers are present in the lower part east of Beer Head.

Bioturbation. The mixing-up of sediment by organisms.

Bivalves. A type of fossil shell with two valves.

Blue Anchor Formation. The highest subdivision of the Mercia Mudstone, present east of Seaton. It consists of 29m of mainly grey and dark grey mudstones.

Blue Lias. A Jurassic formation consisting of mudstone with beds of thin limestone.

Brachiopods. (brack-ee-oh-pods). A type of fossil seashell.

Branchiopod crustacean. A type of crustacean belonging to the Class Branchiopoda.

Breccia(s). (bretch-ee-ah). A type of sedimentary rock consisting of angular rock fragments cemented together.

Budleigh Salterton Pebble Beds. A Triassic formation consisting of about 30m of well-rounded pebbles with some sand beds.

Burrow(s). Tubular holes made by worms, molluscs or other animals, penetrating rock and often filled with clay or sand. A type of trace fossil.

Calcarenite(s). A limestone consisting mainly (more than 50%) of detrital calcite particles of sand size.

Calcareous sandstone. A sandstone cemented by calcite; or a sandstone containing appreciable calcium carbonate but in which detrital quartz is present in excess of 50%.

Calcite. A common mineral, usually white, consisting of calcium carbonate, forming the main mineral in limestones.

Calcrete. A hard mass of calcium carbonate precipitated from solution by infiltrating water in, for example, a sandstone.

Carboniferous. A geological period, between 359.2 and 299.0 million years old. Widespread in central Devon.

Celestine. A mineral, strontium sulphate.

Chalk. A white, very fine-grained type of limestone, of Cretaceous age. Also formations composed mainly of that rock type.

Charmouth Mudstone. Jurassic grey mudstones with few beds of limestone, overlying the Blue Lias.

Chemical [sedimentary rock]. Composed mainly of material deposited by chemical precipitation, whether organic or inorganic. Examples are flint, salt, and gypsum.

Chert. A hard, very finely crystalline silica rock, similar to flint but less dense and less fine-grained. It occurs as layers at some levels in the Upper Greensand, particularly in the Whitecliff Chert.

Clay-with-flints. A residual deposit of reddish brown clay with unworn flints, forming a layer over the Chalk and Upper Greensand. Funnel-shaped solution pipes filled with the deposit are locally present.

Coccoliths. One of the many armour-like plates that surround single-celled marine algae known as coccolithophores.

Conglomerate(s). (kon-glom-er-ayte). A type of coarse-grained sedimentary rock consisting of rounded rock fragments stuck together.

Correlation. The process of establishing the age equivalence of rocks or events in separate areas.

Cotham Member ('Cotham Beds'). The lowermost subdivision of the Lilstock Formation, consisting of greenish grey mudstone and limestone.

Cowstones. Concretionary boulders occurring at some levels in the Foxmould subdivision of the Upper Greensand.

Cretaceous. A geological period covering the time between 145.5 and 65.5 million years ago.

Cross-bedded. Strata that show cross-bedding.

Cross-bedding. A structure in sedimentary rocks in which strata are laid down at an angle to the horizontal. Also known as current bedding or false bedding.

Detrital [sedimentary rock]. Rock formed by the accumulation of minerals and rocks derived from the erosion of previously existing rocks.

Devonian. A geological period, covering the span of time between 416.0 and 359.2 million years ago.

Dunscombe Mudstone Formation. Green and purple mudstones occurring in the middle of the mainly red mudstones of the Mercia Mudstone.

Echinoids. An animal, commonly known as a sea urchin, belonging to the Class Echinoidea, characterised by a roughly subspherical shape with interlocking calcareous plates. Found as fossils particularly in the Chalk.

Escarpment. A long, more or less continuous, steep slope facing in one general direction, usually marking the outcrop of a resistant rock layer in a series of gently dipping softer less resistant strata.

Evaporite. A rock made up of minerals precipitated from solutions concentrated by the evaporation of solvents. Examples are salt, gypsum and anhydrite.

Exe Breccia. The lowest rocks seen in our area, around Exmouth. A type of ancient gravel, consisting of angular fragments of sandstone. Of Permian age.

Fault(s). A fracture in a rock along which there has been some relative movement of the rock on either side. Hence **faulting**.

Feldspar. A very abundant group of rock-forming aluminosilicate minerals.

Flint. A very fine-grained, hard, usually black form of silica, commonly occurring as nodules in the Chalk, and also found reworked into river gravels.

Fold. A curve or bend in rock strata.

Formation. A body of rocks of generally similar type which can be shown on a geological map.

Foxmould. The lowest subdivision of the Upper Greensand in our area, consisting mainly of soft greenish grey glauconitic calcareous sandstone, weathering to 'foxy' brown sand. There is a change in our area from calcareously–cemented beds and concretions in the east to siliceously–cemented horizons in the west.

Freestone. Any stone, especially limestone, that breaks freely and can be cut and dressed with equal ease in any direction without splitting. In our area the Beer Stone is an excellent freestone.

Gastropods. Any mollusc belonging to the Class Gastropoda, characterised by a distinct head with eyes and tentacles and a single calcareous shell that is sometimes spiral, not chambered and generally asymmetrical. e.g. a snail. They occur as fossils in our area especially in the Upper Greensand.

Gault. A patchy thin layer of sandy mudstone which passes up into the Upper Greensand.

Geological time scale. A chronological arrangement of the units of geological time. The time scale used in this book is based on recent research published in 2004 by Gradstein and others (see the Further Reading section).

Glauconite, glauconitic. A name applied to a group of green minerals consisting of hydrous silicates of potassium and iron. It forms on the seabed. Sands containing glauconite have a green colour when unweathered – hence the name greensand, as in the Upper Greensand of our area.

Goyle(s). A local name for a small, deeply incised, stream course.

Gypsum. An evaporite mineral consisting of hydrous calcium sulphate. In our area it occurs in the Mercia Mudstone.

Hardground. A surface, especially in the Chalk and Upper Greensand, representing a small gap in deposition during which the seabed has become hardened. There may also be a concentration of glauconite at the surface.

Head. Locally derived angular rubble formed by downslope movement as a result of freezing and thawing of material in a cold climate.

Honeycomb weathering. A type of weathering in which cavities form on a rock face by chemical or mechanical action, especially along the seashore. Good examples are seen in the Otter Sandstone of our area.

Ice Age. A term used to cover the period of glaciations over the last two million or so years.

Ice sheet. A mass of ice of generally continental size.

Igneous. (ig-knee-us) A type of rock that has solidified from originally molten material.

Impermeable. Said of a rock that does not easily transmit fluids.

Interglacials. The intervals between periods of glaciation.

Joint(s). A fracture in a rock along which no displacement has occurred.

Joint plane(s). The surface of a joint.

Jurassic. A period of geological time spanning the time between 199.6 and 145.5 million years ago. Named after the Jura Mountains between France and Switzerland.

Landslip(s). Areas of soil and rock material which have moved downslope fairly rapidly.

Lilstock Formation. The upper subdivision of the Penarth Group, comprising the Cotham Member overlain by the Langport Member ('White Lias').

Limekiln. A structure in which limestone was burnt to form lime.

Limestone. A sedimentary rock made up mainly of calcium carbonate. There are many different types, some of which contain fossils.

Longshore drift. The process by which material, especially gravel and sand, is moved along the seashore by waves and currents.

Lower Chalk. The traditional lowest subdivision of the Chalk. In our area the Lower Chalk is represented by the Beer Head Limestone.

Marine transgression. The spread of the sea over land areas.

Marl. A clay or mudstone with some content of calcium carbonate, and therefore useful for liming soils.

Marlpit(s). Pits from which marl has been dug for agricultural purposes.

Mercia Mudstone. The group name for a number of geological formations of Triassic age, which are mainly red mudstones, with some grey mudstone near the top. Minor sandstones may also occur.

Metamorphic. A type of rock that has been changed from its original condition by the action of heat or pressure, often deep in the earth's crust.

Middle Chalk. The traditional middle subdivision of the Chalk.

Mudstones(s). A type of sedimentary rock consisting of indurated mud, generally rather blocky and massive in appearance.

New Red Sandstone. A general name for all the mainly red conglomerates, breccias, sandstones and mudstones of Permian and Triassic age in South-west England. Distinguished from the Old Red Sandstone of Devonian age.

Ordovician. A geological period, covering the span of time between 488.3 and 443.7 million years ago.

Otter Sandstone. One of the Triassic formations of our area, occurring between Budleigh Salterton and Sidmouth. It consists mainly of red sandstones with minor conglomerates and mudstones.

Pangaea. A Triassic 'supercontinent' in which all the present landmasses formed one huge landmass.

Penarth Group. Formerly called the 'Rhaetic', it is the topmost geological layer of the Triassic of the Devon coast, it occurs east of Seaton where it overlies the Blue Anchor Formation of the Mercia Mudstone, and consists mainly of dark grey mudstones and grey limestones. It is divided into two formations: Westbury Formation below, and Lilstock Formation above.

Periglacial. Conditions found at the immediate margins of ice sheets, influenced by the cold temperature of the ice.

Periods. In geology, a unit of geological time during which the rocks of the corresponding system, e.g. the Triassic, formed.

Permafrost. Permanently frozen ground occurring in arctic or subarctic regions today, but present in the past also in areas adjacent to ice sheets.

Permian. (Perm-ee-an). A period of geological time, covering the period from 299.0 to 251.0 million years ago.

Playa lake. An ephemeral lake in an arid or semiarid region, that upon evaporation leaves a playa (flat basin floor).

Pleistocene. A subdivision of the Quaternary period, from about 1.8 million years ago to about 10 000 years ago.

Porphyry. Fine-grained igneous rock with conspicuous larger crystals set in the fine groundmass.

Precambrian. The period of geological time from the beginning of the Cambrian 542 million years ago to the formation of the Earth about 4,650 million years ago. It is therefore equivalent to about 90 percent of geological time.

Principle of faunal succession. The principle that groups of organisms have succeeded each other during geological time in a definite order and that each geological period can be recognised by the fossils found in rocks of that age.

Principle of original horizontality. This states that, regardless of the attitude in which we find sedimentary rocks today, they were originally deposited as horizontal layers. There are exceptions where deposition takes place on slopes of up to 30 degrees (e.g. on sand dunes, a channel or a delta).

Principle of the superposition of strata. The principle stating that, in a sequence of layered rocks, the bottom layers were deposited first and are older than those higher in the sequence.

Principle of uniformitarianism. The concept that 'the present is the key to the past'. The principle states that processes now operating to modify the Earth's surface have also operated in the geological past.

Quartz. An extremely common mineral (silicon dioxide), usually colourless or commonly white, but may be many colours.

Quartzite. A type of hard sandstone formed mainly of quartz grains.

Quaternary. The span of time covering the last 1.8 (2.5 according to some authorities) million years.

Radiocarbon dating. A method of dating relatively young (less than 70 000 years old) rocks by measuring the concentration of the radioactive isotope carbon-14 remaining in organic material, for example wood or peat.

Red Rock Gypsum. A distinctive 10m thick layer in the Mercia Mudstone consisting of gypsum veins interwoven with red mudstone.

Reduction spots. Small green spots found in red mudstones such as the Mercia Mudstone.

Rhynchosaurs. A type of Mid Triassic land reptile about 2m long, fossil remains of which are found in the Otter Sandstone.

Ripple marks. Small alternating parallel ridges and hollows found in some sedimentary rocks, especially sandstone, and formed either on land (by wind), or in water (by currents).

River terrace. A flat-topped surface overlying a deposit, usually gravel, and representing the dissected remains of earlier river alluvium.

Salcombe Regis Stone. A type of building stone obtained from the Bindon Sandstone subdivision of the Upper Greensand.

Saltmarsh (deposits). Deposits of salt-marshes, formed in coastal areas subject to occasional flooding by salt water, and consisting mainly of mud and occasional peat.

Sandstone(s). A type of sedimentary rock, composed of sand-sized grains cemented together.

Sea-stack. A small pillar-like rocky island detached from the main coast by wave erosion.

Sedimentary. Said of a type of rock formed of two main types: firstly, made up of detrital fragments worn from pre-existing rocks and laid down either in water or on land, and including mudstones, sandstones, conglomerates and breccias. Secondly, formed by chemical precipitation and including flint and gypsum.

Sedimentation. The process by which sedimentary rocks form.

Sediments. The materials that eventually become sedimentary rocks.

Shales-with-Beef. A subdivision of the Jurassic rocks belonging to the lower part of the Charmouth Mudstone, overlying the Blue Lias around Lyme Regis. It consists of mudstones with numerous thin beds of fibrous calcite or 'beef'.

Shingle ridge(s). A narrow beach of pebbles and cobbles piled up by storm waves.

Shore platform(s). A roughly horizontal surface produced by wave erosion.

Silcrete. A hard conglomerate formed of flints cemented by silica into a hard resistant mass.

Solution pipe(s). A funnel-shaped cylindrical structure extending down into the Chalk or Upper Greensand, and filled with Clay-with-flints.

Strata. See bed(s).

Stratum. Singular of strata.

Submerged forest(s). The remains of an ancient forest now found at low tide or below sea level, indicating a relative rise in sea level since its formation.

Syncline. A fold, which is concave upwards.

Tertiary. The first period of the Cainozoic era, covering the time span from about 65.5 to 1.8 million years ago.

Trace fossils. Structures in sedimentary rocks resulting from the activities of an animal - for example, a track, trail or burrow.

Triassic. A period of geological time covering the period from 251.0 to 199.6 million years ago.

Trilobite(s). A type of extinct marine arthropod belonging to the Class Trilobita, characterised by a three-lobed external skeleton. They lived from Lower Cambrian to Permian times.

Unconformity. A break in a sequence of rocks where one unit of rocks is overlain by another which is not next in order of succession; i.e. there is a substantial time gap between the two sequences. See also angular unconformity.

Upper Chalk. The traditional uppermost subdivision of the Chalk.

Upper Greensand. A formation of Cretaceous age in our area, it consists mainly of sands, sandstones and calcarenites, locally with some beds of chert. It is divided into three; in ascending order, the Foxmould, the Whitecliff Chert and the Bindon Sandstone.

Ventifact(s). A pebble which has been worn, faceted or polished by the sandblasting action of windblown sand in a desert.

Wealden. Rocks of Lower Cretaceous age, mainly mudstones and sandstones, occurring in the Purbeck area.

Westbury Formation. The lowermost of the two formations making up the Penarth Group. It consists of 6 to 8m of dark grey fossiliferous pyritic soft mudstone.

'White Lias'. An earlier term for the unit now called the Langport Member of the Lilstock Formation. It consists of very pale weathering fine-grained limestone, about 9m thick.

Whitecliff Chert. The middle subdivision of the Upper Greensand, consisting of up to up to 32m of sandstones and calcarenites with many layers of nodular and tabular chert.

ACKNOWLEDGEMENTS

I have written most of the book from personal knowledge of the area, but I have also benefited by reading and learning from the work of other writers. It is not possible in a book like this to acknowledge in detail all these authors, but I thank them all for their help. Some of the books and articles that I used are listed in the 'Further Reading' section at the end of the book.

My thanks go to Dr Ramues Gallois who has been most generous in providing me with information and advice in the field and otherwise, and has also kindly provided some excellent photographs and figures for use in the book. Most of the cross-sections are based on his work. He also read and commented on an early draft of the book.

I would also like to thank the following: Dr. Iain Stewart of Plymouth University for his inspirational foreword to the book; Professor Denys Brunsden (Jurassic Coast Trust) and Richard Edmonds (Earth Science Manager, Jurassic Coast) for valuable editorial comments, support and advice; Peter Sills of Coastal Publishing Limited and the financial contributors for supporting publication of the book; Jonathan Lewis (Coastal Publishing Limited) for his design flair; Mark Woods (British Geological Survey) for advising me on updated fossil names; and Bob Symes and John Edwards of Sidmouth Museum for help in locating and providing illustrations.

Thanks to Mark Hounslow of Lancaster University for invaluable advice on the dating of the red rocks based on his research on their palaeomagnetism.

I am grateful to those who have kindly given permission to use their copyright material, as listed in the Image Acknowledgements below.

Finally, I would like particularly to thank my wife Jenny and daughter Katie for their much appreciated advice and encouragement during the writing of the book.

IMAGE ACKNOWLEDGEMENTS

(Key: t:top; m:middle; b:bottom; l:left; r:right; c:centre)

Images in this book are copyright of the photographers and artists.

Photos: Keith Corrie collection, courtesy of Sidmouth Museum: 26bl; Richard Edmonds: 26tr, 40b, 83t, 83m, 103; Dr R W Gallois: 41m, 45m, 91t, 93; Pamela Harrison: 51b; Malcolm Hart, University of Plymouth: 31; Peter Sills: Front cover, 8, 9, 10-11, 12b, 15, 19, 20, 25bl, 28, 30t, 33m 57, 97t, 127; Photo by author, courtesy of Sidmouth Museum: 48; Royal Albert Memorial Museum: 46; Sidmouth Museum: 79t; Simon Powell, University of Bristol: 30b; Ussher Society: 18m; Chris Pamplin: 101m; Beer Quarry Caves: 106-107.

Illustrations: Cross-sections Exmouth-Seaton by the author based mainly on data provided by Dr R W Gallois; Seaton-Lyme Regis section by author. Re-drawn by Jonathan Lewis; Jurassic Coast Trust: 8-9b; Palaeomap Project: 12t, 14m; Sidmouth Museum: 18t; Jonathan Lewis: 14b, 29b; Department of Geology, National Museum of Wales: 16; Courtesy of Bristol University, painted by Pam Baldaro: 35; Reproduced from H B Woodward (1893), *The Jurassic rocks of Britain*, Vol. III: 37; Original by author, re-drawn by Jonathan Lewis; 38t; Drawing reproduced from Buckland, (1823): 38-39; Reproduced from A J Jukes-Browne and W Hill (1900-1903), *The Cretaceous rocks of Britain*, Vols. I-III: 43; Modified and redrawn from chart for Ancient Human Occupation of Britain Project at http://www.nhm.ac.uk/hosted_sites/ahob/AHOBI/overview_time_chart.gif: 47; Original by author, redrawn by Jonathan Lewis: 49; Lyme Regis Museum: 51t; From Roy Millward and Adrian Robinson, The South West Peninsula, 1971, Macmillan Publishers Ltd., reproduced by permission of Palgrave Macmillan: 54; Original by author, re-drawn by Jonathan Lewis: 56; Reproduced from W Vicary and J W Salter (1864) (see this book, p.125): 65t; Jonathan Lewis: 70; Drawn by author based on data provided by Dr R W Gallois, re-drawn by Jonathan Lewis: 83; Dr Peter Grainger: 100; Natural History Museum, London: 101b; Sheila Pollard and Sheila Luxton: 105.

While every effort has been made to contact copyright owners, we apologise if any images have been incorrectly attributed.

FURTHER READING

Arber, M A. 1940. The coastal landslips of south-east Devon. *Proceedings of the Geologists' Association,* Vol. 51, 469-471.

Barber, C, 1993. *Around & about Seaton and Beer.* (Obelisk Publications).

Barber, C. 1996. *Along the Otter.* (Obelisk Publications).

Beche, H T de la. 1839. *Report on the geology of Cornwall, Devon and West Somerset.* Memoir of the Geological Survey of Great Britain. London.

Brown, S and Holbrook, N. 1989. A Roman site at Otterton Point. *Devon Archaeological Society,* Vol. 47, 29-42.

Brunsden, D. (editor). *The official guide to the Jurassic Coast. Dorset and East Devon's World Heritage Coast.* (Coastal Publishing).

Butler, J. 2000. *Travels in Victorian Devon. The illustrated journals and sketchbooks of Peter Orlando Hutchinson, Devon antiquary 1810-1897.* (Devon Books).

Butler, R K. (editor). 1986. *A view from the cliffs.* The East Devon Heritage Coast. (Devon Books).

Campbell, D. 2006. *Exploring the Undercliffs. The Axmouth to Lyme Regis National Nature Reserve.* A 50th Anniversary Guide. (Coastal Publishing).

Cherry, B, and Pevsner, N. 1989. *The Buildings of England. Devon.* Penguin Books.

Conybeare, W D and Buckland, W. 1840. Ten plates....representing the changes produced on the coast of east Devon between Axmouth and Lyme Regis by the subsidence of the land ....on the 26th December 1839, and 3rd February 1840. London.

Cornish, J G. 1933. The Quarries of Salcombe Regis. *Devon and Cornwall Notes and Queries.* Vol. 17, 30-34.

Corrie, D G and Green, D I. 2003. The Barite and supposed celestine from the Triassic mudstone of Sidmouth, Devon. *UK Journal of Mines and Minerals,* No. 23, 27-35.

Dorset Geologists' Association Group. 2003. *Coast and country. Geology walks in and around Dorset.* Dorset Geologists' Association Group.

Durrance, E M and Laming, D J C (editors). 1982. *The geology of Devon.* (University of Exeter).

Edwards, R A and Gallois, R W. 2004. *Geology of the Sidmouth district – a brief explanation of the geological map.* Sheet Explanation of the British Geological Survey. 1:50 000 Sheets 326 and 340 Sidmouth (England and Wales).

Fox, A. 1964. *Ancient peoples and places. South West England.* (London: Thames and Hudson).

Fox, A. 1996. *Prehistoric hillforts in Devon.* (Devon Books).

Gale, A W. 1992. *The building stones of Devon.* Devonshire Association [booklet].

Gallois, R W. 2001. The lithostratigraphy of the Mercia Mudstone Group (mid to late Triassic) of the south Devon coast. *Geoscience in south-west England,* Vol. 10, 195-204.

Gallois, R W. 2003. *Culverhole Point.* Chapter 7 (pp. 55-61) in Coast and Country. Geology Walks in and around Dorset (including excursions within the World Heritage Site). Dorset Geologists' Association Group.

Gallois, R W. and Porter, R J. 2006. The stratigraphy and sedimentology of the Dunscombe Mudstone Formation (late Triassic) of south-west England. *Geoscience in south-west England,* Vol.11, 174-182.

Gibbons, W. 1987. *Scenery and geology around Beer and Seaton.* Holiday Geology Guide [Leaflet].

Gradstein, F M, Ogg, J G, Schmitz, M and Ogg, G. 2012. The geological time scale 2012. (Elsevier).

Grainger, P, Tubb, C D N, and Nielson, A P M. 1985. Landslide activity at the Pinhay water source, Lyme Regis. *Proceedings of the Ussher Society*, Vol. 6, 246-252.

Griffith, F. 1988. *Devon's past: an aerial view.* (Devon Books).

Holmes, G (editor). 1987. *Sidmouth: a history.* (Sidmouth: Sidmouth Museum).

Hoskins, W G. 1972. *Devon.* (Newton Abbot: David & Charles)

Hutchinson, P O. 1843. *The geology of Sidmouth and of South-eastern Devon, illustrated with engravings on copper and wood.* (Sidmouth: John Harvey).

Hutchinson, P O. 1870-1881. *History of Sidmouth.* [Unpublished manuscript. Five quarto volumes, bound in green vellum, are held at the West Country Studies Library, Exeter. A photocopy is available for inspection].

Hutchinson, P O. 1894. *Guide to Sidmouth.* 8th edition. (Sidmouth: Culverwell & Sons).

Jukes-Browne, A J. 1903. *The Cretaceous rocks of Britain: the Lower and Middle Chalk of England.* Memoir of the Geological Survey of Great Britain. London.

Laming, D. 1991. *The Building Stone and its Quarry.* Pages 65-73 in Exeter Cathedral: a celebration. Ed M. Swanton.

Mathews, R. 1998. *It's nice to know Sidbury.*

Millward, R and Robinson, A. 1971. *Somerset and East Devon.* [pp. 169-178. Sidmouth and the Greensand country]. (London: Macmillan).

Myers, B. 1992. *A guide to the River Sid.* (Sid Vale Association).

Otter Valley Association. 1997. *Twelve walks in the Otter Valley.*

Perkins, J W. 1971. *Geology explained in South and East Devon.* (David & Charles).

Pollard, S H M. 1966. Neolithic and Dark Age settlements on High Peak, Sidmouth, Devon. *Proceedings of the Devon Archaeological Society*, Vol. 23, 35-59.

Pollard, S H M. 1967. Radiocarbon dating. Neolithic and Dark Age settlements on High Peak, Sidmouth, Devon. *Proceedings of the Devon Archaeological Society*, Vol. 24, 41.

Risdon, T. 1605-1630 (Published 1811). *Chorographical Description or Survey of the County of Devon.*

Scott, J and Gray, G. not dated. *Out of the darkness.* A brief history and description of the Old Quarry, Beer.

Sid Vale Association. 1996. *Walks on the hills around the Sid Valley. Peak Hill. Salcombe Hill. Fire Beacon Hill.* [Leaflet].

Skertchley, S B J. 1879. On the manufacture of gun-flints, the methods of excavating for flint, the age of Palaeolithic Man, and the connexion between Neolithic art and the gun-flint trade. *Memoir of the Geological Survey.* (London: HMSO).

Vancouver, C. 1808. *Agriculture of Devon.*

Vicary, W, and Salter, J W. 1864. On the pebble bed of Budleigh Salterton with a note on the fossils. *Quarterly Journal of the Geological Society of London*, Vol. 20, 283-302.

Woodward, H B, and Ussher, W A E. 1911. The geology of the country near Sidmouth and Lyme Regis. *Memoir of the Geological Survey.* England and Wales (Second edition).

VISITING THE JURASSIC COAST

The Jurassic Coast is well served by its 'Gateway Towns' and villages which provide a good range of accommodation, museums and visitor centres throughout the year. There are mainline rail links to the area from London and Bristol. Exeter airport is less than half an hour's journey from the coast.

Exploring the World Heritage Coast is best done on foot or by sea. The entire coast is accessible via the South West Coast Path National Trail, which is easy to follow by looking for the acorn symbol. A network of footpaths link to the coast path and make circular walks possible. Full details of access are given in the book. Guided walks and events take place throughout the year and details can be found in Tourist Information Centres and by logging on to www.jurassiccoast.com. All visitor centres and many museums offer regular walks, talks and other events.

There are numerous boat trip operators offering Jurassic Coast tours along the coast and this is really one of the best ways of seeing the World Heritage Site. Tourist Information Centres have more details, and boat trips are also usually advertised around local harbours and quaysides.

Public transport is available between many of the towns and villages along the coast. Please call Traveline on 0871 200 22 33 for specific public transport information.

SAFETY

On the beach and coast path

- Stay away from the base of the cliffs and the cliff top and ensure that children and dogs are kept under control.
- Do not climb the cliffs. Rockfalls can happen at any time.
- Beware of mudslides, especially during or after wet weather.
- Always aim to be on the beaches on a falling tide and beware of the incoming tide, especially around headlands. Be sure to check the tide tables.
- Beware of large waves in rough weather, especially on steeply shelving beaches.
- Observe all permanent and temporary warning signs; they advise on hazards and dangers. Check routes beforehand by visiting www.southwestcoastpath.com
- Be very careful on rocky foreshores which often have slippery boulders.
- Stay within your fitness level – Some stretches of coast can be strenuous and/or remote.
- Make sure you have the right equipment for the conditions, such as good boots, waterproof clothing and sun screen if appropriate.
- Follow The Countryside Code.

Along the undercliff between Axmouth and Lyme Regis

- The walk along the undercliff between Axmouth and Lyme Regis is about 11km and is strenuous.
- There are no 'escape routes' to the beach or main road. Check the map displayed on each board to find out how far you have walked.
- The footpath is often muddy and slippery – stout footwear is essential.
- For your own safety you must keep to the Coast Path. There are deep fissures hidden by dense vegetation.
- Ticks (small sucking parasites) occur here. They can carry Lyme disease, a potentially serious condition. To avoid bites, wear light coloured clothing and long trousers tucked into your socks. Check yourself for ticks. If you are bitten and feel unwell, seek medical advice.

Emergencies

In an emergency dial 999 or 112 and ask for the Coastguard, but be aware that mobile phone coverage in some areas is very limited.

Collecting fossils

- The best, and safest place to look for fossils is on the beach, away from the base of the cliffs, where the sea has washed away soft clay and mud.
- Do not collect from or hammer into the cliffs, fossil features or rocky ledges.
- Keep collecting to a minimum. Avoid removing *in situ* fossils, rocks or minerals.
- The collection of specimens should be restricted to those places where there are plenty of fossils.
- Only collect what you need... leave something for others.
- Never collect from walls or buildings. Take care not to undermine fences, bridges or other structures.
- Be considerate and don't leave a site in an unsightly or dangerous condition.
- Do not use a hammer on flint or chert, which shatter into sharp fragments.
- Some landowners do not wish people to collect – please observe notices.

If you make an important find please contact the nearest visitor centre, or the World Heritage Team on

01305 224132

Visitor Centres and Museums (Some Centres open on a seasonal basis)

Fairlynch Museum, Budleigh Salterton	EX9 6NP	01395 442666
The Arches Jurassic Coast Interpretation Centre	Sidmouth seafront	–
Sidmouth Museum	EX10 8LY	01395 516139
Honiton Museum	EX14 1PG	01404 44966
Fine Foundation Centre	Beer seafront	–
Lyme Regis Philpot Museum	DT7 3QA	01297 443370
Jubilee Pavilion	Lyme Regis seafront	–

Tourist Information Centres (Some Centres open on a seasonal basis)

Exeter	EX1 1GF	01392 665700
Exmouth	EX8 1HL	01395 222299
Budleigh Salterton	EX9 6NG	01395 445275
Ottery St Mary	EX11 1BZ	01404 813964
Sidmouth	EX10 8XR	01395 516441
Honiton	EX14 1LT	01404 43716
Axminster	EX13 5AQ	01297 34386
Seaton	EX12 2TB	01297 21660
Lyme Regis	DT7 3BS	01297 442138

Other Useful Contacts

National Trust	West Dorset	01297 489481
	Devon	01392 881691
South West Coast Path Association		01752 896237
Natural England (Information on National Nature Reserves)		0845 6003078
Geological information	www.dorsetrigs.org.uk	
	www.devon.gov.uk/geology	
Public transport information	www.traveline.info	0871 200 22 33

For other information and a list of useful web sites please visit

www.jurassiccoast.org

Download the FREE Jurassic Coast App. and take a virtual Walk Through Time